Biocultural Diversity Conservation

Biocultural Diversity Conservation

A Global Sourcebook

Luisa Maffi and Ellen Woodley

publishing for a sustainable future
London • Washington, DC

First published in 2010 by Earthscan
Reprinted in 2010

Earthscan Ltd, Dunstan House, 14a St Cross Street, London EC1N 8XA, UK
Earthscan LLC, 1616 P Street, NW, Washington, DC 20036, USA

Earthscan publishes in association with the International Institute for Environment and Development

For more information on Earthscan publications, see www.earthscan.co.uk or write to earthinfo@earthscan.co.uk

ISBN 978-1-84407-921-6 paperback

Typeset by JS Typesetting Ltd, Porthcawl, Mid Glamorgan
Cover design by Clifford Hayes

A catalogue record for this book is available from the British Library

Library of Congress Cataloging-in-Publication Data
Maffi, Luisa.
Biocultural diversity conservation : a global sourcebook / Luisa Maffi and Ellen Woodley.
p. cm.
Includes bibliographical references and index.
ISBN 978-1-84407-920-9 (hardback) -- ISBN 978-1-84407-921-6 (pbk.) 1. Ethnoscience. 2. Biodiversity conservation. 3. Nature--Effect of human beings on. 4. Globalization. I. Woodley, Ellen. II. Title.
GN476.M38 2010
304.2–dc22

2009051712

At Earthscan we strive to minimize our environmental impacts and carbon footprint through reducing waste, recycling and offsetting our CO_2 emissions, including those created through publication of this book. For more details of our environmental policy, see www.earthscan.co.uk.

This book was printed in the UK by MPG Books, an ISO 14001 accredited company.
The paper used is FSC certified.

Mixed Sources
Product group from well-managed forests and other controlled sources
www.fsc.org Cert no. SA-COC-1565
© 1996 Forest Stewardship Council

Contents

Figures, tables, boxes and plates *vii*
Foreword by Gonzalo Oviedo *xi*
Acknowledgements *xiii*
Abbreviations *xv*
Introduction: Why a Sourcebook on Biocultural Diversity? *xix*

Part I – Biocultural Diversity: Conceptual Framework

1 What is Biocultural Diversity? 3

2 Why is a Biocultural Approach Relevant for Sustaining Life in Nature and Culture? 13

Part II – Sustaining Biocultural Diversity: The Projects

3 Surveying Biocultural Diversity Projects around the World 23

4 Overview of the Projects 27

5 Cross-cutting Analysis of the Projects 129

6 Lessons Learned from the Projects 155

Part III – Sustaining Biocultural Diversity: Future Directions

7 Filling the Gaps and Connecting the Dots: Recommendations and Next Steps 175

8 Biocultural Diversity and the Future of Sustainability 191

References 197

Appendix 1: Analytical Tables *209*
Appendix 2: Survey Details *237*
Appendix 3: Survey Contributor Information *243*
Appendix 4: Directory of Selected Resources on Biocultural Diversity *249*
Appendix 5: About Terralingua *267*
Appendix 6: About the Authors *269*

Index *271*

Figures, Tables, Boxes and Plates

Figures

1.1	Relationship between global/regional/national correlations of cultural and biological diversity and causal relationships between cultures and biodiversity at the local level	6
1.2	Endemism in languages and higher vertebrates: Comparison of the top 25 countries	7
2.1	Kayapó People in Pará, Brazil. Humans and nature: separate entities or interdependent whole?	14
2.2	The contribution of traditional farmers to the global stock of plant crop varieties: The central Andes in Peru exhibit the highest inter- and intra-specific agrobiodiversity in the world	16
4.1	The forest–farm boundary at the Kilum-Ijim forest was agreed by participatory decision making with local communities	28
4.2	Community gathering in the Dorbo sacred pasture land to get blessing from indigenous religious leaders (sitting in the front row)	30
4.3	Women displaying *kitete* gourds and *kitete* seed necklaces at a community festival	33
4.4	A women's focus group discussing changes in plant abundance in Goka, Tanzania	36
4.5	Traditional healing centre, Uganda	39
4.6	The woodpile is the Xhosa women's status symbol and cultural totem	41
4.7	Dr Alain Touwaide at work in the Library at Soliman the Magnificent Mosque, Istanbul, examining ancient herbals and documents from which he recovers information about the ancient therapeutic uses of plants	43
4.8	Within one hour, Tibetan villagers in northwest Yunnan collected more than 80 local species, which have been traditionally used for generations and are classified according to their own epistemology and knowledge systems	45
4.9	Peje, an Andamanese, cutting a bamboo for making bows	51
4.10	Agustinus Angkol, a traditional Tado elder and herbarium researcher, showing one of the medicinal plants he collected	55
4.11	Barbara Stuart and Bonnie Marisha Sampi with *Barndu* (water goanna) and *Jalij* (freshwater prawn)	61
4.12	Kinami Mountain on Lihir Island, PNG	68

4.13 Lihir mine site, PNG 69
4.14 Men involved in participatory mapping exercise in Roviana, Solomon Islands 71
4.15 Hills at Tl'oondih where summer and winter trails led to traditional Gwich'in hunting grounds in the Yukon, and a clearing where one Gwich'in elder had his camp 78
4.16 Preparing fish in Pikangikum: people are moulded by the land and everything they draw from it, say the elders 83
4.17 Traditional fishing site in Gitxaała territory showing beach at low tide, with ancient stone fish traps (semi-circular) visible in the intertidal zone 85
4.18 Apache students identifying plants at Goshtlish Tú Bil Sikąné 90
4.19 Luís and Tomás Palma gathering native pine seeds for a tree nursery in their Sierra Tarahumara community 93
4.20 Hotï people drying cane for blowguns 107
4.21 A high level of community participation and capacity building in the development of resource management plans helps foster biodiversity conservation on indigenous lands in Ecuador 110
4.22 Teaching the children in the Upper Amazon region of Peru 112
4.23 Mapping traditional territories in the Xingu Indigenous Park 117
4.24 The Kaxinawá people of Acre, Brazil
4.25 A dance of the Bambuti community of Semliki Forest, Western Uganda 124
4.26 Peruvian biologist María Scurrah learning the names of traditional potato varieties from a farmer in Quilcas, Junín Department, Peru 126
5.1 Commercial logging in the Sierra Tarahumara of northern Mexico is a major source of deforestation, soil erosion and loss of water resources, all of which severely affect local communities 133
5.2 A Tanzanian elder helping with a plant use consensus analysis 143
5.3 Extensive community consultations and an explicit mutual agreement were the preliminaries for collaboration between the Rarámuri of the Sierra Tarahumara, Mexico, and an international NGO 145
5.4 Lydia Ozies in the bush with *mardiwa* (edible gum) in the Kimberley, Western Australia 150
6.1 One of the key challenges faced by the Rarámuri of the Sierra Tarahumara, Mexico is the intergenerational transmission of traditional values, knowledge, practices and language 160
6.2 Mapping northern boreal forests in the Whitefeather Forest Initiative in northern Ontario, Canada, as part of a land use strategy that combines environmental stewardship with economic renewal opportunities for Pikangikum youth 164
7.1 Traditional handcrafts and sustainable livelihoods: A Waorani woman weaving a traditional bag using fibre from the chambira palm 178
7.2 Parents and children in Pitumarca, Cusco, Peru, posing next to a poster that compares traditional and Western forms of knowledge 188
8.1 Chake Chake village, Tanzania. In the perpetuation of biocultural diversity lies hope for the future of sustainability 196

Tables

5.1 Factors affecting biocultural diversity loss 131
A1.1 Overview of the biodiversity and cultural conservation aspects of projects 209
A1.2 Level of participation and means of knowledge transmission in projects; contribution to biocultural diversity policy 225
A3.1 Contact information for project contributors by project name and descriptive title 245

Boxes

1.1 On the importance of language retention 9
5.1 Industrial agricultural and loss of biocultural diversity 134
5.2 Acculturation processes in the Philippines 135
5.3 Countering marine biodiversity loss through reliance on traditional cultural practices 137
5.4 Securing resource tenure for conservation in Latin America 139
5.5 The role of *kitete* landraces in local ceremonies 141
5.6 Collaborative partnerships in the Whitefeather Forest Initiative 146
5.7 The importance of knowledge transmission to youth in Tamil Nadu 147
5.8 Using traditional knowledge for policy decisions 152
5.9 Guardians of the forest 153
6.1 The requirements for fully collaborative projects: A Kenyan perspective 166
7.1 International policies that support biocultural diversity 180
7.2 An indigenous peoples' definition of development 183
8.1 Cultural and ethical values in IUCN's 2009–2012 Programme of Work 194
A2.1 Cover letter to potential sourcebook contributors 239

Plates

1 Global distribution of plant diversity and languages
2 Three core areas of biocultural diversity as identified by the Index of Biocultural Diversity
3 Biocultural diversity at risk: The world's threatened ecoregions and ethnolinguistic groups
4 Map of the world showing the location of the 45 projects in this sourcebook
5 Drawing representing indigenous perspectives of the local environment and visualizing innovative and creative strategies for maintaining the landscape in Yunnan, China
6 Pages from the diary of one of the indigenous Kaxinawá people trained as Agro-Forestry Agents in Brazil
7 Boys in their *chacra* (cultivated field) in Matara, Cajamarca, Peru
8 Girls in the highlands in Cusco, Peru
9 Gamo elders praying in Dorbo Meadow in Southern Ethiopia at the beginning of the Mascal ceremony
10 Documenting the morphological diversity of *kitete* gourds in Kitui District of Kenya

11 Young Agta girl spearfishing in the Disulap River Philippine crocodile sanctuary in the municipality of San Mariano, Isabela Province, Luzon, Philippines
12 Learning by doing is an important method of knowledge transmission among the Eñepa people of Venezuela

Foreword

Gonzalo Oviedo

The timing of the publication of this book couldn't be more appropriate, as we celebrate in 2010 the International Year of Biodiversity. There are always very good reasons to celebrate biodiversity, but there are equally good reasons to worry, as the 2010 target is proving elusive – 'to achieve by 2010 a significant reduction of the current rate of biodiversity loss at the global, regional and national level as a contribution to poverty alleviation and to the benefit of all life on Earth'. At the same time, when the UN reviews progress in September 2010 towards the MDGs and other international development goals, there will not be many indications of success, and thus not many reasons to celebrate.

Perhaps the topics addressed in this book will offer some clues as to why such fundamental objectives of the international community seem so difficult to meet. As it is known to biologists, diversity contributes to ecosystems' resilience – and there are growing indications that the same applies to human cultures. As the prevailing economic models and political systems continue to promote standardized, homogeneous responses to the needs and challenges of development and conservation, we lose diversity. We also lose resilience, as many people find themselves increasingly alienated from their cultural strengths – the knowledge and practices for survival and adaptation accumulated through generations. Policies and practices that better understand the profound links between nature and culture, and the value of diversity for resilience, can support creativity, encourage better-adapted responses and empower people to value their identity and knowledge.

This book represents a culmination of the already extensive contributions that the authors have made towards meeting these objectives. From presenting the conceptual issues in a solid but accessible manner, to researching examples of biocultural practices worldwide, to extracting lessons that others can benefit from, their work is filling a critical gap of knowledge and policy. The last decade has seen a growing interest at the international level in the links between biological and cultural diversity, as shown in a large number of events, research projects and papers; equally important, although perhaps less visible, has been the strengthening and multiplication of grassroots experiences that recuperate and invigorate traditions and that reinforce local cultures and institutions, including through political processes. But what has been missing is a robust articulation of the conceptual approaches, in ways that make justice to the richness of local expressions. This Reader is a very valuable contribution to filling this gap.

As noted by the authors, much remains to be done, despite the progress made. A particular challenge that deserves to be highlighted is developing and applying a biocultural approach to climate change adaptation. There are many paradoxes in climate change, and one of them is that, grossly speaking, vulnerability of human groups is inversely proportional to their responsibility in generating or aggravating climate change; and vulnerability is closely connected with cultural diversity – that is, a large part of the most vulnerable populations of the rural world is made up of indigenous and traditional groups that are experiencing severe impacts, while having limited means, opportunities and support to overcome such impacts.

The other paradox, of course, is that such cultures have long histories of adaptation – indeed, many of their cultural features are basically the result of adaptive responses to climate variability and associated ecosystem change. But successful adaptation requires not only traditional knowledge and skills – it also demands capacity to control and confront the key drivers of vulnerability, which most such cultures lack in this world of marginalization and inequity. Here, therefore, is the challenge – to promote adaptation approaches that build on the adaptive traditions and the resilient cultural institutions, and strengthen the social fabrics of the communities, while reducing the risks associated with marginalization and poverty.

As this sourcebook points out, resilience is a central concept in rethinking sustainability; resilience not only of ecological but also of social systems, and the interconnectedness of both. There is no sustainable future without greater resilience at all those levels. As we move towards new ways of framing our responses to the challenges we face, the lessons and reflections of this book will be a fundamental reference.

Gonzalo Oviedo
Senior Adviser on Social Policy
International Union for Conservation of Nature (IUCN)

Acknowledgements

Our first thanks go to Jim Nations (then at Conservation International), for prompting one of us (Maffi) to develop a series of case studies on an integrated approach to sustaining cultures and biodiversity, as a way to further promote understanding, appreciation and adoption of a biocultural perspective on the diversity of life on Earth.

We are deeply grateful to The Christensen Fund for providing the necessary funding, through two consecutive grants (2003–2008), to the organization that spearheaded the project, the NGO Terralingua (see Appendix 5). Without this generous support, the project could not have seen the light and come to fruition.

We thank Rebecca Stranberg, a graduate student at the University of Guelph (Ontario, Canada), where the project was first located in 2003–2004, for setting up and handling our electronic database, conducting web and library searches, and helping develop the survey tools. Preston Hardison (Cultural Stories Project, Tulalip Tribes), developer of the ICONS software and database, graciously made ICONS available to us for use in our project. Philip Cottrell of Ripplemusic in Guelph, Ontario, was responsible for the initial development of a discussion forum located on Terralingua's website and meant to foster communication among the project's leaders and contributors.

We also benefited greatly from the comments of several reviewers. Dr Alan Hedley, Professor Emeritus at the University of Victoria (British Columbia, Canada) and earlier on the governing board of Terralingua, acted as an internal reviewer and critic. His careful attention to our conceptual framework, methodology and analysis contributed significantly to improving the rigour of our presentation of the materials. The comments of three anonymous external reviewers prompted us to fill important gaps in information and to strengthen our arguments about the relevance of a biocultural approach to conservation.

Terralingua staff members Tania Aguila and Ortixia Dilts were closely involved in the later stages of the project and in the finalization of the manuscript. Tania helped to edit and format several versions of the text and ensure that the manuscript was ready for submission. Ortixia prepared the book's illustrations according to the technical requirements, contributed to the design of the book's cover, and compiled the lists of illustrations, tables and boxes. She has also developed the dedicated biocultural diversity conservation portal on Terralingua's website that provides the online companion to this book. Both Tania and Ortixia offered helpful comments and suggestions on the book's form and content. Their support and dedication were much appreciated.

Above all, we are greatly indebted to all those around the world who first took the time to respond to our sourcebook survey and send information about their projects, and then patiently fielded additional requests for details, feedback on draft project descriptions, project photos, interesting project 'tidbits', and more. Needless to say, without their contributions this project would not have taken shape. Their names appear at the beginning of the respective project descriptions, but we wish to acknowledge and thank them all here for their interest, input and patience through the lengthy process of development of the sourcebook.

These people are: Anvita Abbi, Shankar Aswani, William Balée, Jonathan Barnard, Chad Kālepa Baybayan, Nathan Cardinal, Andrew Chapeskie, Michelle Cocks, Darron Collins, John DeMarco, Desalegn Desissa, Sarah Edwards, Simon Foale, Renato Gavazzi, Yogesh Gokhale, Hugh Govan with Rigoberto Carrera, Jorge Ishizawa with Grimaldo Rengifo, Xu Jianchu, Kimberley Language Resource Centre Aboriginal Corporation (contact person: Siobhan Casson), Ingrid Kritsch, Jonathan Long, Martha Macintyre, C. Manjula, Dawn Marsden, Fulvio Mazzocchi, Charles Menzies, Jonathan Miller, Felipe Montoya Greenheck, Yasuyuki Morimoto, Gabriel Nemogá with Carlos Mamanché, Laxmi Pant, Giulia Pedone, Alex Peters, Jeanine Pfeiffer with the Tado and Waerebo Communities and Elizabeth Gish, Jan van der Ploeg, João de Queiroz, David Rapport, Samantha Ross, Sekagya Yahaya Hill, Alain Touwaide, David Turnbull, Nancy Vander Velde with Jorelik Tibon, Helen Verran, Patricia Vickers, Bruce White, Rob Wild, Zerihun Woldu, Stanford Zent. Janelle White provided information on one of the projects early on. (Full contact information for project contributors is found in Appendix 3.)

Abbreviations

ACT	Amazon Conservation Team
AMAAI-Ac	Associação do Movimento dos Agentes Agroflorestais Indígenas do Acre (Brazil)
AMNH	American Museum of Natural History
ARC	Aboriginal Rainforest Council Inc.
ARC	Australian Research Council
ATIX	Indigenous Land Association of the Xingu (Brazil)
ATK	aboriginal traditional knowledge
BC ACADRE	British Columbia Aboriginal Capacity and Developmental Research Environment
CAIMAN	Conservación en Áreas Indígenas Manejadas (Ecuador)
CAR	Regional Autonomous Corporation (Columbia)
CBD	Convention on Biological Diversity
CBE	community-based ecotourism
CBMPA	Community-Based Marine Protected Area
CBO	community-based organization
CEESP	Commission on Environmental, Economic and Social Policy (IUCN)
CEM	Commission on Ecosystem Management (IUCN)
CFC	Couples for Christ (Solomon Islands)
CIHR	Canadian Institute of Health Research
COSEWIC	Committee on the Status of Endangered Wildlife in Canada
CPI/Ac	Commisão Pró-Indio do Acre (Brazil)
CSIN	Canadian Sustainability Indicators Network
CSVPA	Cultural and Spiritual Values of Protected Areas (IUCN)
EARTh	Environmental Applications Reference Thesaurus project
ECO-SEA	Ethnobotanical Conservation Organization for Southeast Asia
FAO	Food and Agriculture Organization (UN)
FMI	forest management institution
FUNAI	National Indian Foundation (Brazil)
GBCDA	Global Biocultural Diversity Assessment (Terralingua)
GIS	geographic information systems
GSCI	Gwich'in Social and Cultural Institute
GSR	Gwich'in Settlement Region
GTZ	Gesellschaft für Technische Zusammenarbeit
IAFAs	Indigenous Agro-Forestry Agents (Brazil)

IBCD	Index of Biocultural Diversity
ICCA	Indigenous and Community Conserved Area
ICE	Indigenous Cooperative on the Environment
IDRC	International Development and Research Centre
IGO	intergovernmental organization
IITC	International Indian Treaty Council
IK	indigenous knowledge
ILD	Index of Linguistic Diversity
IPO	indigenous peoples' organization
ISE	International Society of Ethnobiology
IUCN	International Union for the Conservation of Nature
IVIC	Instituto Venezolano de Investigaciones Científicas
KAWG	Kyanika Adult Women's Group
KLRC	Kimberley Language Resource Centre
LGL	Lihir Gold Limited
LGU	Local Government Unit
LINKS	Local and Indigenous Knowledge Systems in a Global Society (UNESCO)
LK	local knowledge
LRMP	Land Resource Management Planning
MAB	Man and the Biosphere (UNESCO)
MPA	Marine Protected Area
NACAs	Nuclei for Andean Cultural Affirmation (Peru)
NGO	non-governmental organization
NHS	National Historic Site (Canada)
NRM	natural resource management
PBR	People's Biodiversity Register (India)
PNG	Papua New Guinea
PRATEC	Proyecto Andino para las Tecnologías Campesinas (Peru)
RCF	Roviana Conservation Foundation
RMI	Republic of the Marshall Islands
SARA	Species at Risk Act (Canada)
SCB	Society for Conservation Biology
SSWG	social science working group (SCB)
TAMI	Text, Audio, Movies and Images (Northern Australia)
TCC	Theme on Culture and Conservation (IUCN/CEESP)
TEK	traditional environmental knowledge
TILCEPA	Theme on Indigenous and Local Communities, Equity, and Protected Areas (IUCN)
UBC	University of British Columbia
UNAS	Universidad Nacional Agraria de la Selva (Peru)
UNDP	United Nations Development Programme
UNEP	United Nations Environment Programme
UNESCO	United Nations Educational, Scientific and Cultural Organisation
UNPFII	United Nations Permanent Forum on Indigenous Issues
USAID	United States Agency for International Development
VITEK	Vitality Index of Traditional Environmental Knowledge

WCC	World Conservation Congress (IUCN)
WCPA	World Commission on Protected Areas (IUCN)
WHC	World Heritage Convention (UNESCO)
WiSATA	West Manggarai Swiss Australia Tourism Assistance
WSCP	Western Solomons Conservation Program
WSSD	World Summit on Sustainable Development
WTWHA	Wet Tropics World Heritage Area

Introduction: Why a Sourcebook on Biocultural Diversity?

Luisa Maffi

Biodiversity also incorporates human cultural diversity, which can be affected by the same drivers as biodiversity, and which has impacts on the diversity of genes, other species, and ecosystems. (UNEP, 2007, p160)

In October 1996, a small group of researchers, practitioners and activists – from the natural and the social sciences, and from indigenous and non-indigenous backgrounds – gathered in Berkeley, California (US) for a small working conference, titled 'Endangered Languages, Endangered Knowledge, Endangered Environments'. They met to discuss the links between linguistic, cultural and biological diversity and the threats shared by these diversities. At that time, probably none of the participants would have imagined that, a decade later, one might read a statement such as the one quoted above, from the flagship report on the state of the global environment issued periodically by the United Nations Environment Programme (UNEP). The belief in the interconnectedness of humans and nature has been widespread in the worldviews of many indigenous and traditional societies. However, the idea that the diversity of life on Earth is biological, cultural and linguistic diversity – or 'biocultural diversity' for short – was still novel and poorly understood in academic circles, much less enshrined in policy documents. Among the few precedents was a pioneering statement about the existence of an 'inextricable link' between cultural and biological diversity contained in the Declaration of Belém, issued by the International Society of Ethnobiology in 1988, which was one of the inspirations for the 1996 Berkeley conference.

The distance covered since then is remarkable. The concept of biocultural diversity has become an object of academic enquiry, with lines of research covering topics such as: GIS (geographic information systems)-based studies of the overlapping global and regional distributions of biodiversity and cultural diversity and the biophysical and social factors accounting for the patterns observed; the development of quantitative methodologies for the measurement and assessment of the state and trends of biocultural diversity at global and regional scales; and on-the-ground investigations of the co-evolved relationships between cultures and ecosystems at the local level. The relevance of this concept for biodiversity conservation, sustainable development and the deployment of the human potential has begun to penetrate the international agenda, from UNEP to the Convention on Biological Diversity (CBD), the International Union for the Conservation of Nature (IUCN) and the United Nations Educational, Scientific and Cultural Organisation (UNESCO), to name a few. Clearly, the idea of biocultural diversity is coming of age.

Yet, there is no question that much more needs to be done. Although countless grassroots efforts are underway worldwide to sustain cultures and biodiversity in an integrated fashion, most of these efforts fall 'under the radar' and lack the ability to connect with other similar endeavours elsewhere. This significantly limits their ability to form a united front and achieve greater visibility. As a consequence, the lessons from these activities remain dispersed in many different locales and cannot be learned easily. Their wide-ranging implications for policy and implementation – and indeed for an overall paradigm shift in how we think of human relationships with the environment – cannot be brought out as prominently as they deserve. That is where this sourcebook comes in: to 'connect the dots' among a meaningful selection of such efforts and make the lessons learned widely available.

About this book

This volume is the first resource of its kind, meant to serve as a synthetic and informative reference on biocultural diversity conservation for researchers, professionals, policy makers, indigenous and other local organizations, international agencies and non-governmental organizations (NGOs), funders, media and others. The book is the outcome of a project carried out over several years by Terralingua (www.terralingua.org; see Appendix 5): the Global Sourcebook on Biocultural Diversity. The material presented here is based on a worldwide survey that we conducted beginning in 2004. Our goal was to identify projects that take an integrated, synergistic approach to sustaining local cultures and biodiversity: that is, *projects that emphasize the close integration of biodiversity conservation with the maintenance and revitalization of cultural and linguistic heritage*. In other words, we sought projects that recognize the fundamental link between local language, ecological knowledge, cultural practices and biodiversity, and that apply this recognition to the design of sustainable solutions to environmental and social problems. We were especially interested in projects initiated and conducted by indigenous and local communities, or else jointly planned, led and managed by these communities and external agents (such as governments, international organizations or NGOs).

It is important to clarify at the outset that we did not mean to carry out a 'scientific' survey – that is, one based on the kind of rigorous sampling methodology that characterizes, for example, sociological surveys. This would have been both beyond our means in undertaking the work and beside our point in charting the emerging field of biocultural diversity conservation. Nor did we intend to conduct a systematic comparison of the merits or demerits, successes or failures, of biocultural approaches to conservation versus non-biocultural ones. Such an endeavour would be very worthwhile, but it was beyond the scope of our work. It would also have been premature, since as yet there has been insufficient understanding of the nature and dynamics of biocultural projects as such. Achieving the latter kind of understanding was therefore our main goal in carrying out this project. First and foremost, we aimed to understand what works where, when and how in biocultural diversity conservation and what improvements can be made; to foster experience sharing and mutual learning among those who are involved in applying the concept of biocultural diversity to on-the-ground action; and to ensure that the lessons would be accessible to a wider audience.

The survey yielded 45 biocultural projects, programmes and initiatives from all continents (see Plate 4). The description and analysis of these activities and the discussion of 'lessons learned' from them form the core of this sourcebook, and are meant to offer guidance for future efforts to sustain and restore the world's biocultural diversity. The projects we present here well illustrate the remarkable variety of activities that are being undertaken around the globe to sustain and restore biocultural diversity. At the same time, these 45 projects are undoubtedly just the 'tip of the iceberg' of the integrative biocultural work that is taking place worldwide. For this reason, we produced a companion portal to this sourcebook, hosted on Terralingua's website at www.terralingua.org/bcdconservation. The portal includes a database of biocultural diversity projects that can be expanded and updated over time. The portal also hosts a discussion group specifically devoted to the exchange of ideas among sourcebook contributors and others interested in biocultural diversity conservation. The formation of a network of like-minded individuals actively involved in supporting biocultural diversity is helping to identify the needs and requirements for promoting bioculturally oriented research and action and for advancing shared goals. Such a 'community of practice' will significantly advance the further promotion and development of the biocultural approach.

The answer to the question: 'Why a sourcebook on biocultural diversity?' ultimately resides in making the broadest possible impact beyond the circle of the 'already converted'. By increasing the visibility of integrated biocultural endeavours vis-à-vis policy makers, international agencies and NGOs, funders, media and others, this sourcebook aims not only to benefit indigenous and local communities, researchers and professionals who are involved in such efforts, but also to affect thinking on a global scale. This is an ever more pressing goal at a time in which – over 20 years after the Brundtland Commission's report *Our Common Future* (WCED, 1987) – we are inescapably confronted with the unsustainability of a dominant model that places an exponentially increasing strain on the natural fabric of our planet and dramatically erodes biodiversity and the health of the world's ecosystems (Millennium Ecosystem Assessment, 2005; UNEP, 2007; WWF, ZSL and GFN, 2008). In this context, the inextricable link between cultural diversity and biodiversity and the importance of cultural diversity for sustaining the diversity of life must be high on the international agenda. The global community is beginning to take stock – witness for example the inclusion of an extensive set of biocultural diversity events under the banner of a 'Biocultural Diversity and Indigenous Peoples Journey' in the Conservation Forum at IUCN's Fourth World Conservation Congress in Barcelona, Spain, in October 2008, and the repeated mention of the importance of cultural diversity for biodiversity and sustainability in IUCN's 2009–2012 programme of work. This volume is intended as a contribution to fostering this process, through which the concept and practice of biocultural diversity conservation are becoming mainstream.

Organization of the book

Following this introduction, the book is organized in three parts. Part I, 'Biocultural Diversity: Conceptual Framework', defines biocultural diversity, introduces a 'conceptual map' for the links between cultures and biodiversity at different scales, and synthesizes research advances in this field (Chapter 1). It then discusses the relevance of the biocultural approach to maintaining and restoring the diversity of life in nature and culture (Chapter 2).

Part II, 'Sustaining Biocultural Diversity: The Projects', describes the survey's criteria, areas of emphasis and process (Chapter 3) and provides a description of each project, programme or initiative surveyed, cross-referenced to a world map showing the projects' locations (Chapter 4). The following chapter presents a cross-cutting analysis of all projects around several sets of criteria, based on both the original survey materials and the extensive follow-ups with the contributors (Chapter 5). This analysis is cross-referenced to two analytical tables included in Appendix 1, which provide a convenient synopsis of the project by the main criteria we used in information gathering and analysis. Chapter 6 then turns to lessons learned from the projects, as relevant to biocultural diversity conservation.

Part III, 'Sustaining Biocultural Diversity: Future Directions', identifies existing gaps in research, practice, policy and education; points to avenues for further development; and formulates a set of recommendations for researchers, practitioners, international agencies, policy makers and an extended circle of interested people and organizations (Chapter 7). The last chapter (Chapter 8) provides a final synthesis and reviews the place of the concept and practice of biocultural diversity within the broader context of the ecological and social sustainability of our planet.

The Appendixes, in addition to the synoptic tables mentioned above (Appendix 1), present the survey methodology and tools (Appendix 2), the survey contributor information (Appendix 3), a directory of other useful resources on biocultural diversity (Appendix 4), information about Terralingua (Appendix 5) and information about the authors (Appendix 6).

Part I

Biocultural Diversity: Conceptual Framework

1

What is Biocultural Diversity?

Luisa Maffi

The case studies presented in this volume are an eloquent illustration of 'biocultural diversity in practice': the on-the-ground application of the idea that maintaining and restoring the diversity of life means sustaining both biodiversity and cultures, because the two are interrelated and mutually supportive. Whether or not the projects, programmes and initiatives surveyed here make this assumption explicitly, they are all, in different forms and to different degrees, informed and guided by this basic idea. It is a testimony to the strength of the idea that its application does not necessarily follow from a single, unified conceptual framework (that is, as if it were a test of a theory), but rather seems to emerge spontaneously, and to some extent independently, time and time again in different places. This is, arguably, the hallmark of ideas whose time has come: they build from the ground up, and from many different sources.

Indeed, this is the case with the idea of 'biocultural diversity', an idea that has emerged at the intersection of different disciplines and knowledge systems – from anthropology to linguistics, ethnobiology, ethnoecology, conservation biology, ecology, and indigenous knowledge – and that has drawn significantly from on-the-ground experiences worldwide. In many ways, this idea is still emerging and evolving as a multifaceted and 'organic' concept, true to the nature of the reality it seeks to describe. Nevertheless, over the past decade some recurring elements have become apparent in the work of those who have been more closely concerned with this idea, so that it is now possible to outline some basic elements and definitions. Some key areas of enquiry have also taken shape, from global and regional mappings and analyses of the distribution patterns of biocultural diversity, to the study of the links between biodiversity and culture at the local level, to the development of methodologies and tools for assessing and monitoring the state and trends of biocultural diversity worldwide. Here, we review these theoretical and methodological advances in order to offer the readers a synthetic background on the emerging field of biocultural diversity and a framework for the case study presentation and analysis that forms the core of this volume.

Defining biocultural diversity

Like other species, humans are an intrinsic part of the natural environment. Throughout the history of our species, humans have always made use of and modified the natural environment in response to their material and non-material needs. At the same time, human cultures have adapted to the natural environment in which they have developed, and thus have been influenced and shaped by this adaptation process. Cultural beliefs, values, institutions, knowledge systems, languages and practices manifest this mutual relationship between humans and the environment: they both express this relationship and are the means through which this relationship has been formed. The diversity of the world's cultural systems (or 'cultural diversity' for short) envelops the globe, forming what some have conceptualized as a 'logosphere' – a planetary web of human languages (Krauss, 1996) – and others as an 'ethnosphere' – a planetary web of human cultures (Davis, 2001). Both concepts are reminiscent, to some extent, of the much earlier notion of 'noosphere', the planetary web of human cognition proposed by Teilhard de Chardin (1966).

This complex system of cultural diversity does not simply parallel the diversity found in the natural world; it is profoundly interrelated with it (Posey, 1999; Maffi, 1998, 2001, 2005, 2007a; Harmon, 2002; Stepp et al, 2002; Carlson and Maffi, 2004; Maffi and Woodley, 2007; Kassam 2009). The organization, vitality, and resilience of ecosystems and those of human communities are mutually linked (Berkes and Folke, 1998; Rapport, 2007; Rapport and Maffi, 2010). All humans are immersed in this web of interdependence, no matter how close or remote their daily contact with the natural world may be. The current state and the future of human societies are inextricably tied to those of the natural environments in which people live. However, perception of this link is frequently weaker in industrialized, urbanized societies, where people inhabit built environments and are removed from a direct dependence on nature for their subsistence. Awareness of the link remains stronger in indigenous or local communities that maintain direct ties to and immediate dependence on their natural environments. A view of humans as part of, rather than separate from, the natural world is in fact pervasive in indigenous societies, and so is a felt connection between language, cultural identity and land (Blythe and McKenna Brown, 2004).

It is becoming increasingly apparent that the breakdown of the perceived link between humans and nature underlies many of the environmental and social problems humanity faces today. The historic loss of understanding of the finiteness and fragility of the natural world, which has come with urbanization and industrialization, goes along with the ability, brought about by economic globalization, to turn a blind eye to the profound social and environmental consequences of massive exploitation and transformation of nature. Together, these two factors create a deleterious feedback loop that further pushes our planet – and humanity with it – toward the brink. Thus, there is an ever more pressing need to understand the connections between biodiversity and cultural diversity, and to act on this understanding, in both policy and practice, to support and restore vitality and resilience to biocultural systems.

Since the early 1990s, interest in these linkages has led to the development of a new field of research and action, which centres on the notion of 'biocultural diversity' (Dasmann, 1991; Nietschmann, 1992; McNeely and Keeton, 1995; Mühlhäusler, 1995; Nabhan, 1997; Nabhan et al, 2002; Posey, 1999; Maffi, 1998, 2001, 2005, 2007a; Harmon, 2002; Stepp et al, 2002; Carlson and Maffi, 2004; Maffi and Woodley, 2007; Kassam, 2009; Rapport and Maffi, 2010). Based on how it is generally understood among biocultural diversity researchers, this concept may be defined as follows (slightly modified from Maffi, 2007a):

> *Biocultural diversity comprises the diversity of life in all of its manifestations – biological, cultural, and linguistic – which are interrelated (and likely co-evolved) within a complex socio-ecological adaptive system.*

This definition comprises the following key elements:

- The diversity of life is made up not only of the diversity of plants and animal species, habitats and ecosystems found on the planet, but also of the diversity of human cultures and languages.

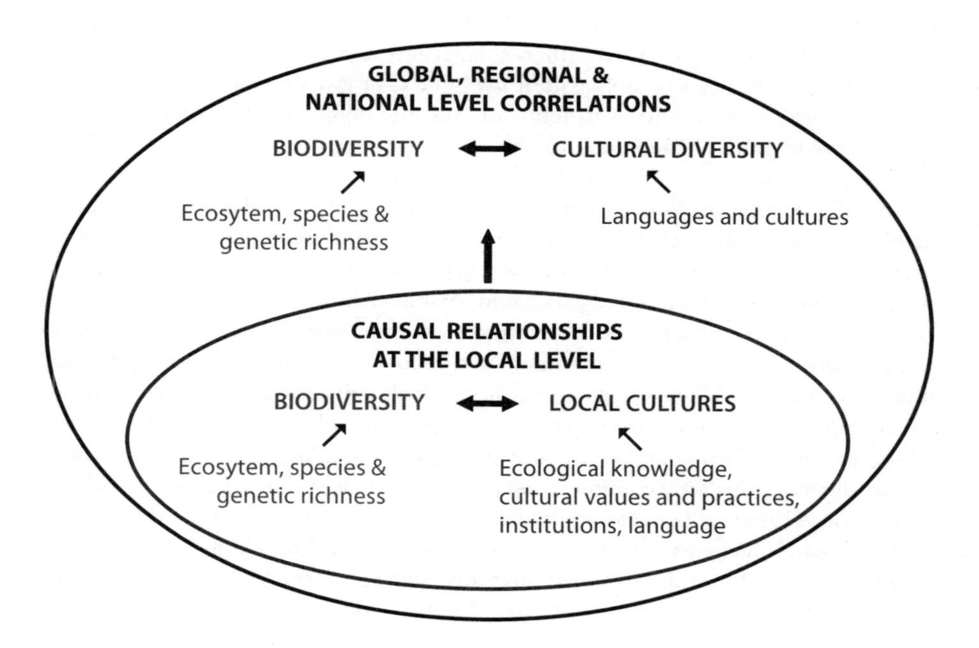

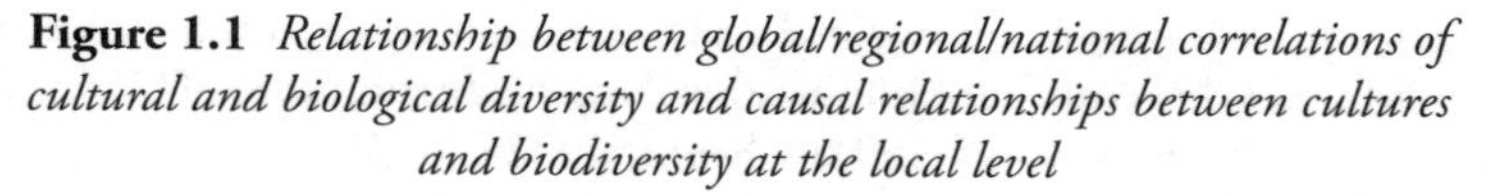
Figure 1.1 *Relationship between global/regional/national correlations of cultural and biological diversity and causal relationships between cultures and biodiversity at the local level*

Source: Original work by Ellen Woodley for Terralingua; modified from Maffi (2007a)

- These diversities do not exist in separate and parallel realms, but rather are different manifestations of a single, complex whole.
- The links among these diversities have developed over time through the cumulative global effects of mutual adaptations, probably of a co-evolutionary nature, between humans and the environment at the local level.

These complex relationships are represented in Figure 1.1.

Understanding biocultural diversity: Global and regional correlations

Recent research has explored these relationships at different scales. Cross-mapping the global distribution of biodiversity and that of cultural diversity has revealed significant overlaps in the respective geographic patterns, especially in tropical areas (Harmon 1996; Maffi, 1998; Oviedo et al, 2000; Skutnabb-Kangas et al, 2003; Stepp et al, 2004). This point is illustrated by Plate 1, which shows the overlapping distributions of the world's plant diversity zones and of the world's languages (the latter used as a proxy for cultural diversity).

These studies also point to a strong correlation between biological and linguistic 'megadiversity' in individual countries (Harmon, 1996). Figure 1.2 presents this correlation, focusing on the overlap in the distribution of endemic languages and of endemic higher vertebrate species (that is, languages and species exclusively found in specific regions, in this case only within the borders of given countries). Many of the top 25 'megadiverse' countries in terms of endemic high vertebrate species are also megadiverse in terms of endemic languages.

Similar findings result from studies that have sought to develop integrated measures for the joint assessment of the state of global biodiversity and cultural diversity. The Index of Biocultural Diversity (IBCD) (Harmon and Loh, 2004; Loh and Harmon, 2005) aggregates selected measures of cultural diversity (numbers of languages, ethnicities and religions) and biodiversity (numbers of bird, mammal and plant species) to provide a country-by-country assessment of the state of biocultural diversity. The IBCD has three components: a 'biocultural diversity richness' component, which is the sheer aggregated measure of a country's richness in cultural and biological diversity; an 'areal' component, which adjusts the indicators for a country's land area, and thus measures biocultural diversity relative to the country's physical extent; and a 'population' component, which adjusts the indicators for a country's human population, and thus measures biocultural diversity in relation to a country's population size. For each country, the overall IBCD then aggregates the figures for these three components, yielding a global picture of the state of biocultural diversity in which three areas emerge as 'core areas' of exceptionally high biocultural diversity: the Amazon Basin, Central Africa and Indomalaysia/Melanesia (see Plate 2).

Research at both global and regional scales has identified a number of geographic and climatic factors that correlate with these overlapping distributions of biodiversity and cultural diversity (Nichols, 1990, 1992; Chapin, 1992 [2003]; Mace and Pagel, 1995; Wilcox and Duin, 1995; Harmon, 1996; Nettle, 1996, 1998, 1999; Lizarralde, 2001; Smith, 2001; Collard and Foley, 2002; Moore et al, 2002; Manne, 2003; Sutherland,

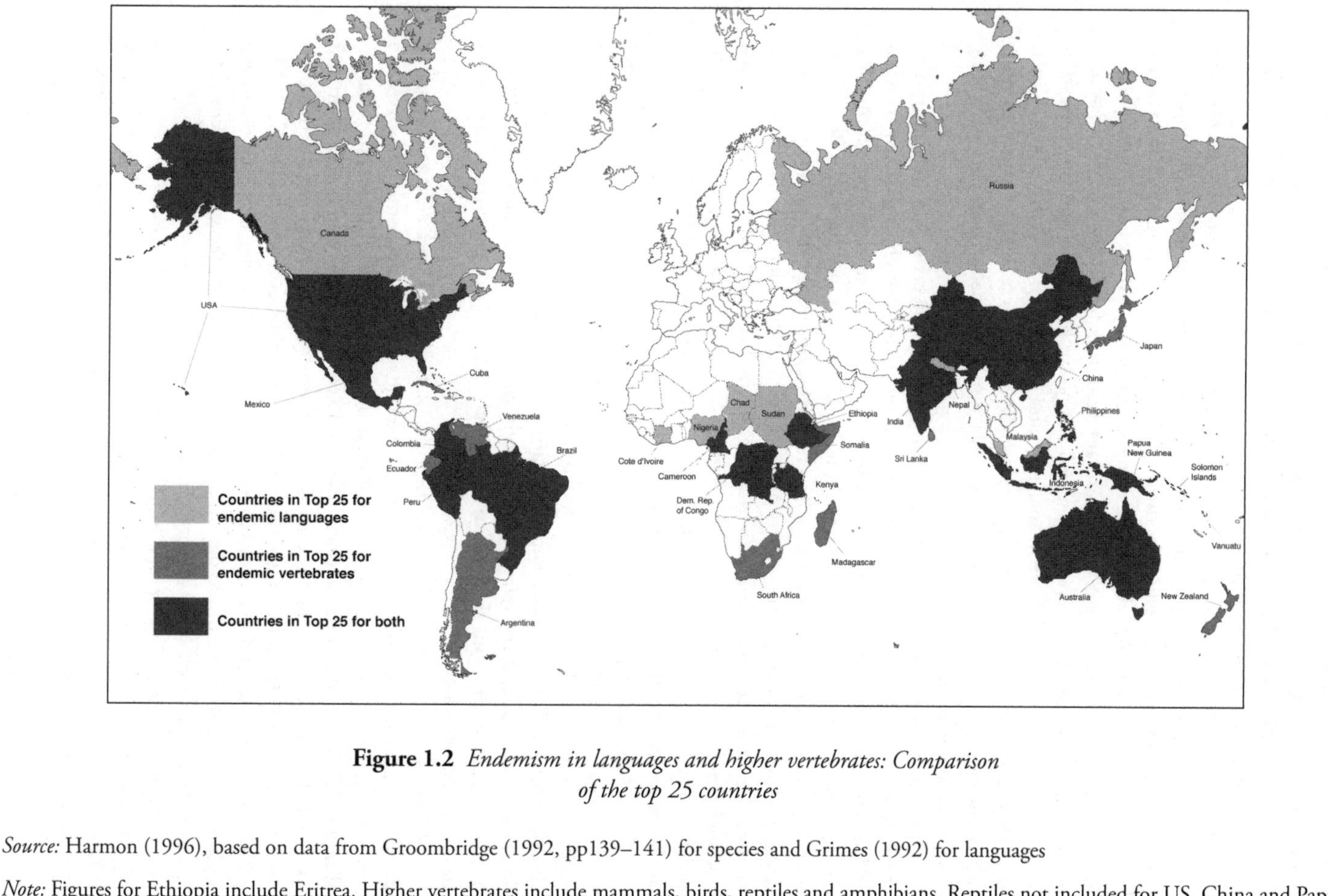

Figure 1.2 *Endemism in languages and higher vertebrates: Comparison of the top 25 countries*

Source: Harmon (1996), based on data from Groombridge (1992, pp139–141) for species and Grimes (1992) for languages

Note: Figures for Ethiopia include Eritrea. Higher vertebrates include mammals, birds, reptiles and amphibians. Reptiles not included for US, China and Papua New Guinea because the numbers were not reported in the source table.

2003; Stepp et al, 2004, 2005). Lower latitudes, higher rainfall, higher temperatures, coastlines and high altitudes positively correlate with both high linguistic diversity and high biological diversity. Higher latitudes, plains and drier climates tend to correlate with lower diversity in both realms.

These studies have also identified a variety of social factors that may account for some of the patterns observed. One of these factors is the difference in modes of subsistence (localized vs. wide-ranging), which in turn is influenced by how geography and climate differentially affect both the abundance of resources in a given area and human access to these resources. Ease of access to abundant resources found locally (for instance, in tropical forests) seems to favour localized boundary formation and diversification into larger numbers of small human societies (and languages). Resource scarcity (such as in deserts or tundras) and the necessity to have access to a larger territory to meet subsistence needs appear to favour diversification into smaller numbers of widely distributed populations (and languages).

Instead, a lowering of both cultural and biological diversity has been found to correlate with the development of complex, stratified and densely populated societies and of far-reaching economic powers. Although specific historical and biophysical circumstances must be considered in seeking to account for any such case, it appears that high population densities may correlate with domination by a single or a few ethnic groups, with detrimental effects on local cultural diversity and the surrounding biodiversity. From ancient empires to today's globalized economy, these complex social systems have spread and expanded well beyond the confines of local ecosystems, exploiting and draining natural resources on a large scale and imposing cultural assimilation and the homogenization of cultural diversity.

These findings raise important questions of history, pattern, causality, scale and levels of analysis. How have the links among diversities developed and changed over time, how are these relationships manifested today, and how does one form of diversity affect the others? Is local biodiversity, at least to some extent, a cultural product? How do these diversities and the relationships among them present themselves at different degrees of resolution, from the local to the global, and how are patterns and processes connected across scales?

Whereas correlations in the distribution of biological and cultural-linguistic diversity appear strong at the global level, analyses at smaller scales sometimes present a mixed picture in terms of the patterning of these diversities (Smith, 2001). For instance, although Central and South America, West and Central Africa, South and Southeast Asia and the Pacific stand out as areas of high biocultural diversity, such correlations may weaken when 'zooming in' on these regions at higher degrees of resolution. This difference in outcome at different scales stresses the need for further studies of the global and regional distributions of biological, cultural and linguistic diversity, both currently and over time. Research is now being carried out with these more complex questions in mind (for example, Stepp et al, 2005, 2008). As this line of research evolves, we can expect to gain a much deeper understanding of the geography of biocultural diversity at global and regional scales, as well as the factors involved in the persistence or loss of biocultural diversity in various parts of the world.

Biodiversity and culture at the local level

While studies at global, regional, and national scales bring out the *correlations* between biological and cultural diversity, detailed case studies at the local level are needed to understand the *causal links* between the environment and cultural values, beliefs, institutions, knowledge systems, practices and languages, and the changes that affect the persistence or loss of these links. Ethnobiologists and ecological anthropologists who have documented traditional environmental knowledge (TEK) have pointed to the value of TEK for the well-being and livelihoods of local communities, sustainable resource use, environmental conservation, and the analysis and monitoring of long-term ecological changes (Berlin, 1992; Williams and Baines, 1993; Balée, 1994; Berkes, 1999; Gragson and Blount, 1999; Medin and Atran, 1999; Posey, 1999; Minnis and Elisens, 2000; Maffi, 2001; Krupnik and Jolly, 2002; Stepp et al, 2002; Carlson and Maffi, 2004; Ellen, 2006; Kassam, 2009). Social, cultural, political and economic transformations have profound impacts on TEK and its links with local ecosystems and biodiversity. Such changes commonly include external exploitation of traditional lands and resources or loss of tenure over such lands and resources, displacement, out-migration, impoverishment, forced or induced assimilation, loss of cultural identity and acculturation to a dominant way of life, shift from local languages to majority languages, integration into a market economy, and loss of local decision-making capacity and self-sufficiency. These processes often also bring about a dramatic deterioration in the affected people's physical, psychological, social and spiritual well-being. As an example, in a public opinion poll conducted in Canada, 63 per cent of First Nations respondents identified the loss of land and culture as significant contributors to poorer health status (UNPFII, 2006). Complementarily, another Canadian study (Chandler and Lalonde,

Box 1.1 On the Importance of Language Retention

In the same way that a healthy planet requires biological diversity, a healthy cultural world requires linguistic diversity. Yet, language is also an elaborate phenomenon tied to real people and cultures. Language loss threatens a fundamental human right – that of expression of the life and life ways of a people.

Each language relates ideas that can be expressed in that language and no other. Thus, when an indigenous community is no longer allowed to pray, sing or tell stories in its language, it is denied a fundamental human right. Unfortunately, linguistic rights have been seriously abused for hundreds of years by banning specific languages and indirectly by assaulting language-support structures such as land, economies and religions.

Languages today are the next frontier in setting the country into moral and environmental symmetry.

Source: Meya (2006)

1998) showed that health and well-being are highest in those First Nation communities that retain their language and culture or have taken steps to rebuild cultural continuity on their lands. Wilhelm Meya, Director of the Lakota Language Consortium in the US, clearly points to these connections (and their implications for human rights), in a statement (Meya, 2006) that refers specifically to the US but is applicable worldwide (Box 1.1).

The progressive erosion of the diversity of traditional knowledge and value systems also represents a global loss: a depletion of the pool of adaptive solutions developed by humans worldwide in addressing social and environmental problems – and thus a diminished potential for future adaptations. It is therefore crucial to have systematic means for assessing the state of traditional knowledge. Very relevant in this connection is some of the recent quantitative research carried out by ethnobiologists to measure the retention and erosion of TEK. Researchers such as Zent (1999, 2001), Lizarralde (2001), Ross (2002), Zarger and Stepp (2004), Zent and López-Zent (2004), and others have contributed to the development of quantitative methods for analysing the acquisition and intergenerational transmission of ethnobotanical and ethnoecological knowledge, and for identifying factors (such as age, formal education, bilingual ability, language shift, length of residency, change in subsistence practices and so forth) that may influence the maintenance or loss of TEK. This research is essential for understanding the links between the retention or erosion of TEK and a variety of both ecological and social factors, including changes in natural resource use and management that can affect the state of biodiversity and ecosystems. For example, a body of TEK accumulated over generations and attuned to local ecological circumstances can be rendered irrelevant if social changes force indigenous and local communities to adopt unsustainable practices in relation to the environment (Hunn, 1999).

Because language is a fundamental means of communication, transmission and storage of knowledge and values, it is also essential to understand the factors that positively or negatively affect the vitality of local languages. An expert group on language endangerment and language maintenance (UNESCO, 2003) has put forth a set of recommendations for the assessment of linguistic vitality. These experts point to a variety of 'vital statistics' needed for this purpose, such as: numbers of mother-tongue speakers over time, intergenerational language transmission, contexts of use, availability of mother-tongue education and so forth. Researchers have recently developed a methodology for testing linguistic vitality at the local level and identifying the factors (such as age, gender, special roles and so forth) that influence linguistic ability in situations of cultural and linguistic change (Florey, 2006). This work importantly complements the quantitative tools for measuring TEK retention and loss. Overall, this research on TEK and linguistic vitality significantly improves our ability to explore the causal links between language, knowledge and the environment.

Trends in biocultural diversity

A crucial step in understanding these links at the local level and the correlations between biodiversity and cultural diversity at global and regional scales is being able to address the question: Do the trends of global biodiversity parallel those of cultural diversity? If it could be shown that they do, this would provide further support for the interrelatedness of these diversities. Again taking languages as proxies for cultural diversity as a whole, if languages are being lost, and if language loss is a factor in the erosion of cultural values, knowledge and practices relevant to the environment, then a reduction in linguistic and cultural diversity could significantly affect the state of biodiversity. Conversely, a downward trend in biodiversity could have an especially negative impact on indigenous and local communities directly dependent on local ecosystems.

Data on biodiversity provide ample evidence of persistent downward trends (Millennium Ecosystem Assessment, 2005; UNEP, 2007; WWF, ZSL and GFN, 2008). Further, a growing number of reports indicate that the world's languages, knowledge systems and cultural traditions are also seriously at risk. Linguists' estimates have suggested that over 50 per cent of the world's approximately 7000 languages are currently endangered (Wurm, 2001), and some researchers have predicted that up to 90 per cent of existing languages may not survive beyond 2100 (Krauss, 1992). Mappings of threatened ecosystems and threatened languages show overlapping patterns in their respective distributions (see Plate 3).

In spite of the urgency of addressing these issues, until recently there was a dearth of time-series data at any scale, from local to global, that would allow for the systematic tracking of trends in linguistic and cultural diversity, and reveal whether these trends mirror those in biodiversity. New work (Harmon and Loh, 2009) has begun to provide trend data on linguistic diversity through the development of an Index of Linguistic Diversity (ILD). The ILD measures changes in the numbers of mother-tongue speakers of the world's languages over time (one of the 'vital statistics' called for by the UNESCO experts). This index will allow for the systematic monitoring of trends in linguistic diversity in the same way that it is currently possible to monitor trends in biological diversity. In turn, Zent (2008) has developed the methodology for the first systematic, fully replicable, locally relevant and globally applicable tool for measuring trends in persistence or loss of TEK: the Vitality Index of Traditional Environmental Knowledge (VITEK). When fully deployed, the VITEK will provide trend data on a key aspect of cultural diversity.

In addition to addressing fundamental questions in the field of biocultural diversity, these tools will be useful to local communities for their self-assessment, and also have the potential to affect policy making and help direct on-the-ground conservation and revitalization efforts. For example, wherever trend data should indicate that biocultural diversity is diminishing, such a finding will call for urgent remedial action to counter these negative trends; conversely, wherever biocultural diversity should appear to be resilient, continual monitoring will help ensure that it continues to thrive.

2

Why is a Biocultural Approach Relevant for Sustaining Life in Nature and Culture?

Luisa Maffi

As the previous chapter should have made readily apparent, biocultural diversity research has significant implications for both biodiversity conservation and the maintenance of cultural vitality and resilience. The case studies presented and analysed in Part 2 provide concrete evidence of the variety of integrated biocultural approaches that people and organizations around the world are developing to address the challenges of sustaining life in nature and culture. As the body of biocultural research and applications grows and becomes more visible, academic institutions, international agencies and governmental and non-governmental organizations have begun to take notice. Nevertheless, the concept and practice of biocultural diversity have yet to gain mainstream recognition and acceptance. Before turning to our case studies, it may be useful to explore some of the obstacles that have stood in the way, and to review some of the evidence and debates that have sought to move the agenda forward.

Views of humans and nature as separate entities

A long-held and widespread view in Western philosophical thought depicts humans not as part of nature, but as separate from it and meant to be dominant over it (Eldredge, 1995). Historically, the biological sciences have tended to reflect this view, seeing nature as exclusively moulded by biological evolutionary processes, and as existing in a 'pristine' state, unless and until humans encroach upon it for purposes of development and natural resource exploitation.

Figure 2.1 *Kayapó People in Pará, Brazil. Humans and nature: separate entities or interdependent whole?*

Credit: Cristina Mittermeier

This pervasive view of nature has had profound implications for mainstream approaches to nature conservation. Conservation professionals trained in the natural sciences have mostly focused on the role of humans in threatening biodiversity and ecosystems through the ever-escalating pace and scope of resource extraction and transformation of nature. Until recently, there was little or no awareness of how humans have also positively contributed to shaping the natural world, in the course of long-term adaptive interactions with their environments. It has been common among conservationists to define and represent ecosystems and ecological processes 'humans out' – that is, to seek to reconstruct an idealized state of ecosystems prior to any human presence, or at least prior to intensified human impact on the environment. Such an idealized picture has been used as a baseline for developing conservation visions and to build benchmarks for measuring success in reaching conservation targets (Mittermeier et al, 1998; Olson and Dinerstein, 1998). 'Conservation and Development' efforts have often been decried as flawed in their attempt to combine what was seen as the incompatible goals of biodiversity conservation and human development in ecologically sensitive areas such as tropical forests (Terborgh, 1999; Schwartzman et al, 2000a; Soulé, 2000). The prevailing 'humans out' conceptualization of nature has also affected conservationists' thinking in relation to parks and protected areas, whose traditional human residents have frequently been seen as a threat to the last remaining 'pristine' environments on the planet (Terborgh, 1999). From this standpoint, it followed that protecting these areas meant excluding the traditional residents from them – while, in many cases, admitting visiting tourists in (but see Harmon, 1998; Brown et al, 2005; for critiques of the exclusionary approach to conservation).

It is undeniable that the exponential increase in the pace and scale of human activities has come to constitute the prime threat to the environment – both through the direct effects of resource extraction and use, and through the indirect results of these activities (such as global climate change). It is now widely recognized that we have entered an era in which human action is causing massive species extinctions, habitat deterioration and loss of ecosystem functions, and that in turn these changes are severely threatening human well-being (Millennium Ecosystem Assessment, 2005; UNEP, 2007; WWF, ZSL and GFN, 2008). Scientists have in fact proposed the term 'Anthropocene' for the most recent era in Earth's history, one characterized by a major human impact on the planet's climate and ecosystems (Crutzen and Stoermer, 2000). Recent work (Ellis and Ramankutty, 2008) has even redefined and mapped the world's biomes in terms of anthropogenic patterns, based on the identification of sustained, direct human interactions with ecosystems. However, it is one thing to recognize the extent of anthropogenic impact on the environment; it is altogether another to paint all of humanity with the same brush. Doing so obscures the fact that human relationships with the environment are a highly complex and diverse phenomenon (Callicott, 1994; Selin, 2003), and that they should be understood on the basis of a wide range of social, cultural, economic, political and ecological variables (Posey, 1999; Maffi, 2001; Harmon, 2002).

Human–environment interactions: Anthropological perspectives

As we discussed in Chapter 1, anthropological and other social science research provides data for this kind of more nuanced perspective on human–environment interactions. This research suggests that small-scale societies with a history of long-term (and unchallenged; see Nietschmann, 1992) occupation of given territories tend to develop and maintain in-depth and accurate knowledge about the local ecology and biodiversity (Berlin, 1992; Hunn, 1999; Shepard, 2004). In many such instances, there is also evidence of low-impact practices of use and management of natural resources, maintained over long periods of time with no detrimental effects on biodiversity and ecosystem functions. In fact, such practices have often been found to contribute to sustaining and even enhancing biodiversity and ecosystem functions, while introducing subtle modifications that mimic natural processes (Posey and Balée, 1989; Williams and Baines, 1993; Berkes, 1999; Hunn, 1999; Carlson and Maffi, 2004). In some cases, research has shown that major ecosystems such as tropical forests, commonly thought of as the quintessential 'pristine' environments, actually bear the mark of vast anthropogenic alterations brought about by resident indigenous populations over long periods of time (Heckenberger et al, 2003, 2007). Small-scale human communities have been identified as 'agents of creative ecological disturbance' (López-Zent and Zent, 2004) and even as 'keystone societies' (Meilleur, 1994).

Among the roles of humans as biodiversity-enhancing agents, Zent and López-Zent (2007) have identified the following:

Figure 2.2 *The contribution of traditional farmers to the global stock of plant crop varieties: The central Andes in Peru exhibit the highest inter- and intra-specific agrobiodiversity in the world*

Credit: Jorge Ishizawa

- the anthropogenic creation and maintenance of biodiverse landscapes through traditional low-impact resource management practices (Posey, 1984, 1998; Denevan and Padoch, 1987; Baleé, 1993; López-Zent and Zent, 2004);
- the contribution of traditional farmers to the global stock of plant crop varieties (Brush, 1980; Boster, 1984; Oldfield and Alcorn, 1987; Thrupp, 1998);
- the customary beliefs and behaviours that contribute directly or indirectly to biodiversity conservation, such as sustainable resource extraction techniques, sacred groves, ritual regulation of resource harvests and buffer zone maintenance (Moock and Rhoades, 1992; Posey, 1999);
- the dependence of the socio-cultural integrity and survival of local communities on access to traditional territories, habitats and resources (Maffi, 2001; Baranyi and Weitzner, 2006).

Sustainability of human–environment relationships: Debates

While at present it is clear that the global impact of human action on the environment is unsustainable, this imbalance has developed over a long period of time, mostly related to the rise and spread of complex civilizations that grew beyond the limits of local ecosystems and began to extend their political and economic reach to draw resources from their 'periphery' (Eldredge, 1995; Wright, 2004; Diamond, 2005). Such processes have often had a profound impact on small-scale societies, imposing rapid and sweeping socio-economic, political and cultural changes that have drastically affected the ways of life, livelihoods and well-being of these societies. While people naturally react to changing circumstances in an attempt to adapt and continue to develop, often one of the most far-reaching consequences of fast and radical change is a loss of control over the process of change itself (Bodley, 1990; Posey, 1999). Such change also has the potential to provoke a major shift in how the people affected perceive their relationship with the environment. Often, this may lead to the weakening of long-held holistic views of people as part of nature, and of nature as a life-giving force to be respected, while new perceptions of 'natural resources' as commodities to be exploited and less sustainable relationships with the environment may begin to take hold (Aumeeruddy, 1994; Kellert et al, 2000; Nations, 2001).

Biodiversity loss and ecosystem degradation in small-scale societies, both historically and at present, have also been attributed to migration to new territories, where lack of familiarity with the local ecosystems would have unwittingly resulted in rapid resource depletion. The case of the extinction of megafauna in North America and the Pacific Islands is often mentioned as an example of destructive exploitation of the environment by migrating societies, although debate rages as to whether climate shifts, or a combination of climate change and over-hunting, may instead have been the main cause (Burney and Flannery, 2005; Robinson et al, 2005; Wroe et al, 2006). In other cases, environmental collapse in sedentary societies has been ascribed to a form of societal 'implosion', such as in the oft-quoted example of Easter Island, where internal conflict would have led to a disastrous escalation in the overuse of natural resources (Diamond, 2005). On the other hand, research also reveals instances in which, over time, human groups that had migrated to a new environment developed local knowledge and ultimately established a dynamic balance with their natural surroundings; and even cases of complex societies that were able to maintain an ecological balance for long periods of time (Diamond, 1991; Ponting, 1991; Bahn and Flenley, 1992; Eldredge, 1995; Flannery, 1995; Kirch and Hunt, 1996; Atran and Medin, 1997; Kirch, 1997; Diamond, 2005).

Another much-debated issue has been the question of whether indigenous peoples are 'conservationists', and whether any 'conservationist' or conservation-like behaviour is the rule or rather the exception among indigenous peoples (Johnson, 1989; Hames, 1991; Alcorn, 1993, 1996; Redford and Stearman, 1993; Ellen, 1994). Implicitly or explicitly, the terms of this debate have been: Do indigenous peoples display conscious conservation-oriented behaviour? Or is it rather that biodiversity conservation in indigenous territories has been largely the unintended result of low population densities, low-impact technologies and long-term adaptation to natural cycles for resource extraction, combined with the indirect effects of cultural sanctions (such as taboos or other societal prescription and proscriptions)? Some critics have pointed to overly romanticized portraits of indigenous and local communities as wise 'stewards of nature', arguing that this misrepresentation has produced a backlash on several levels (Brosius, 1997). In some instances, this backlash has led to charges that the construal of indigenous peoples as 'ecologically noble savages' obscures the occurrence of instances in which indigenous peoples' practices may have a role in environmental degradation, such as with so-called 'slash and burn' agriculture in the tropics (Redford, 1990). Others have suggested that this construal has engendered misplaced expectations that indigenous and local communities should behave like idealized 'museum specimens' in their relationships with the environment – an assumption that negates their right and ability to adapt and develop in response to changing circumstances (Hyndman, 1994). Even in those cases in which indigenous and minority groups themselves have consciously adopted the language of stewardship in an effort to assert their rights, they have sometimes been described as engaging in 'ecopolitics' and 'strategic essentialism' (Conklin and Graham, 1995; Kuper, 2003; Kenrick and Lewis, 2004).

This is certainly not to say that there are no examples of effective collaborations between indigenous peoples and conservationists to sustain biodiversity in indigenous territories and protected areas (for instance, Stevens, 1997; Beltran, 2000; Weber et al, 2000). However, issues related to indigenous peoples and conservation have often led to polarized positions, as for instance in the pages of *Conservation Biology* (Chicchón, 2000; Marsh et al, 2000; Redford and Sanderson, 2000; Schwartzman et al, 2000a, 2000b; Terborgh, 2000), or in some recent critiques of conservationists' approaches toward indigenous peoples (Chapin, 2004; Colchester, 2004; Dowie, 2005; but see Romero and Andrade, 2004 for an attempt to mediate between 'preservationist' and 'devolutionist' perspectives).

A biocultural approach to sustaining life

From a biocultural diversity perspective, the debate about 'indigenous (and local) peoples as conservationists' is ill-formulated. For human behaviours to lead to conservation or sustainable use of the environment and biodiversity, it is not necessary that they be guided by explicit theories comparable to those underlying fields of academic science, such as conservation biology. Some of the scepticism about the value of indigenous and local knowledge for conservation may stem from a misconceived expectation of this sort. Many traditional beliefs and values may indeed be stated in explicit form to consciously guide conservation or 'wise use' behaviour, although they may be conceptualized and expressed in terms of ethical principles (such as not taking more than one needs or thinking of descendants seven generations down the line), rather than formalized as 'conservation guidelines'. In many other cases, however, conservation-like behaviour may arise implicitly from what Atran (2001) calls an 'emergent knowledge structure' – a fluid theory-like belief system that takes shape through cultural upbringing and allows both for informal learning from others and for independent observation of the natural world.

Hardison (2005, pp44–45) points out that, in this sense, 'working with traditional knowledge is less an issue of "integrating" the Western science and traditional knowledge by finding an algorithm to map one system into the other', than it is a matter of acknowledging and respecting cross-cultural differences in knowledge systems and seeking common ground. What matters the most from a biocultural perspective is the very diversity of adaptive tools deployed by human societies in relation to the environment, and the continued intergenerational development, transmission and vitality of beliefs, values, institutions, knowledge, languages and practices relevant to human–environment relationships. From this, it follows that the main goal of a biocultural approach to sustaining life in nature and culture is to understand and support these adaptive tools, as well as the ability for these tools to develop from within their cultural context when new circumstances arise that require new adaptations.

In this light, debating whether or not indigenous peoples and local communities are 'conservationists' (in terms recognizable to biologists or other academically trained scientists and practitioners) is not only misconceived; it also alienates and adds to the disenfranchisement and marginalization of those who have already been most vulnerable to social, political, economic and environmental problems that are often beyond their responsibility and control. A more appropriate consideration is how indigenous peoples and local communities can better engage in and benefit from conservation policies and projects for and on their lands – both by resorting to traditionally well-adapted knowledge systems and resource use and management practices, and by adopting new, explicit conservation measures that may, if appropriate, incorporate elements of formal science (Shepard, 2002).

Supporting biological, cultural and linguistic diversity globally may also be the best chance for all of humanity to 'keep our options alive' as a species (Hunn, 1999). By and large, cultural evolution may have overtaken biological evolution in *Homo sapiens*, but our species still partakes of the fundamental characteristic of the evolution of life: the tendency toward increasing diversification in the deployment of life's potential (Harmon, 2002). The global forces that are reducing this potential in the realm of culture – by threatening the survival of the many thousands of indigenous and local societies that represent the vast majority of the world's cultural diversity – are increasing the likelihood that convergence toward dominant cultural models will cause more and more people to encounter what linguist Peter Mühlhäusler (1995) has called the same 'cultural blind spots': undetected instances in which the prevailing cultural models fail to provide viable solutions to societal and environmental problems. Instead, Mühlhäusler suggests, 'it is by pooling the resources of many understandings that more reliable knowledge can arise'; and 'access to these perspectives is best gained through a diversity of languages' (Mühlhäusler, 1995, p160). In this sense, languages can be understood as a 'resource for nature' (Maffi, 1998), and it is possible to argue that 'any reduction of language diversity diminishes the adaptive strength of our species because it lowers the pool of knowledge from which we can draw' (Bernard, 1992, p82; also see Fishman, 1982; Diamond, 1993). In short, as Pattanayak (1988, p380) puts it, 'ecology shows that a variety of forms is a prerequisite for biological survival. Monocultures are vulnerable and easily destroyed. Plurality in human ecology functions in the same way.'

There are signs that these calls for sustaining cultural diversity along with, and in its interaction with, biodiversity are beginning to be heeded in the conservation world (Redford and Brosius, 2006). A meeting entirely devoted to this topic, the symposium 'Sustaining Cultural and Biological Diversity in a Rapidly Changing World: Lessons for Public Policy', co-organized by the American Museum of Natural History (AMNH), the Theme on Culture and Conservation of IUCN's Commission on Environmental,

Economic, and Social Policy, and Terralingua, was held in April 2008 at AMNH in New York. Further, as we mentioned in the Introduction, biocultural diversity featured prominently in the Conservation Forum at IUCN's Fourth World Conservation Congress, in October 2008.

While this perspective is slowly becoming mainstream at the international level, innumerable activities are taking place worldwide at the local level, with the aim of strengthening the retention and intergenerational transmission of cultural values, beliefs, institutions, knowledge, languages and practices, and of bringing these to bear on the solution of local environmental and social problems. Some of these activities are entirely the initiative of indigenous and local communities, with little or no external support – and often prove to be some of the most integrative efforts to maintain and restore cultural resilience, linguistic vitality and biodiversity. In other cases, activities are initiated by outsiders in collaboration with indigenous and local communities, or invited by these communities as partnerships with outsiders. Due to their very nature, in most instances these local efforts tend to be isolated, and their visibility is low. More often than not, local communities do not have the means and infrastructure, or sometimes even the felt need and desire, to publicize their activities through the communication channels that are commonly used in the 'global village' – from glossy brochures to flashy websites. Furthermore, even many of the larger – national or international – organizations that engage in conservation and sustainable development activities with local communities do not always have established and reliable mechanisms for institutional learning through systematic recording and archiving of such local experiences.

This limitation inevitably affects many of the efforts of direct interest here, that is, the kind of integrated biocultural projects and initiatives that are the object of this volume. This makes it more difficult for their important lessons to be disseminated and shared, so as to have a bearing on policy and implementation at national, regional and international levels. It also reduces the ability of these experiences to have an impact on that vast portion of humanity that seems to have lost awareness of the inextricable link between biological and cultural diversity. What is needed is a way to 'connect the dots' among these initiatives, in order to increase their collective visibility, bring out the voices of people on the ground, and thus strengthen the impact of biocultural approaches. This is the central goal that this sourcebook seeks to accomplish, through the case studies presented and analysed in the following chapters.

Part II

Sustaining Biocultural Diversity: The Projects

3

Surveying Biocultural Diversity Projects around the World

Luisa Maffi and Ellen Woodley

In this chapter, we describe the process through which we carried out our survey of biocultural diversity conservation projects, programmes and initiatives worldwide. We first present the criteria that guided our selection of projects and the areas of emphasis on which we chose to focus. We then briefly outline the data-gathering activities we carried out during the two phases of the survey and the data analysis we conducted in the following stage. The technical aspects of the survey procedures, as well as the survey tools, are provided in Appendix 2.

Project selection criteria

In order to conduct a systematic survey of biocultural diversity projects, we developed a set of guidelines for project selection. According to our operational criteria, projects suitable for our purposes would be ones with the following characteristics:

1 *Being integrative and synergistic.* We were interested in projects that specifically emphasize the *integration of biodiversity conservation and the maintenance or revitalization of cultural (including linguistic) diversity, and the synergies between the two.* In other words, projects exclusively devoted to one or the other of these two aspects would fall outside the realm of our survey; so would, in principle, projects of the kind commonly labelled 'integrated conservation and development', as they tend to focus on socio-economic development, without significant consideration of the cultural aspects that were a key element in our survey. Rather, the projects we sought would recognize the essential connections and interdependence between the

environment and local cultural values, beliefs, institutions, knowledge, practices and languages, and would build on these linkages to address environmental and social problems in an integrated and synergistic fashion. As an example, a project that would seek to reinforce specific cultural practices that are beneficial for the conservation of local biodiversity – such as traditional prohibitions with respect to the use of certain areas of forest, or customs that result in sustainable harvesting of natural resources – would count as an *integrative* project according to this criterion. Such a project would also be *synergistic* to the extent that, in recognizing and valuing cultural practices that help conserve biodiversity, it would provide a greater incentive for using those practices in a purposeful and directed way.

2 *Recognizing the importance of intergenerational transmission of local cultural values, beliefs, institutions, knowledge, practices and languages*. Ideally, projects suitable for our survey would involve *active support for the continued intergenerational transmission of cultural traditions and languages, recognized as crucial to sustainable human–environment relationships*, rather than focus on documenting threatened or endangered linguistic and cultural heritage for the purpose of salvaging it for posterity. While fully appreciating the value of the latter kind of endeavour, in principle we sought projects that would work to ensure the continued vitality and resilience of local cultural values, beliefs, institutions, knowledge, practices and languages by supporting cultural dynamics and institutions that maintain the connection between generations and the flow of information across them.

3 *Being endogenous or strongly participatory*. Desirable projects for our survey were ones *initiated and conducted by local people, or else jointly planned, led and managed by local people and outsiders in a genuinely collaborative manner, with a specific aim to address the needs of local communities*. In establishing this criterion, we wanted to draw a distinction vis-à-vis projects that are not fully collaborative with local communities: in particular, projects that are explicitly 'extractive' in nature (that is, projects that gather resources and information locally and take them ex-situ, without seeking to benefit local communities); or projects that might at best be 'consultative' in nature (that is, in-situ projects that do not directly involve local people, although they may occasionally use information from local sources for their purposes).

Setting these operational parameters was necessary for the purposes of defining the scope of our survey. At the same time, even within these parameters we expected to find significant variation in the approaches taken by the projects we would encounter. Some might more closely approximate an idealized notion of an 'integrative biocultural project', while others might emphasize certain aspects over others. Some projects, for example, might have biodiversity conservation as their main goal, but in that context acknowledge the importance of local cultures; others might focus on language documentation, while recognizing the links between language and local biodiversity. We felt that it would be desirable to make room for a wide spectrum of projects within our general parameters, as there would be much to be learned from the very diversity of their approaches. Therefore, in reviewing the responses to our survey we chose to be fairly inclusive, thus departing from the idea, which has widespread currency in this kind of exercise, of identifying 'best practices'. While the three criteria above broadly guided our selection, we avoided screening the projects we surveyed against one pre-defined model of what a biocultural diversity conservation project 'ought' to be like.

Areas of emphasis

In addition to our general selection criteria, we identified four specific *areas of emphasis* that, based on our understanding of the dynamics of biocultural diversity, we considered relevant for supporting cultural vitality and resilience. We examined whether and how the projects we surveyed focused on the following dimensions:

1. *Cultural practices that conserve biodiversity*: Focus on local beliefs, practices and innovations that, intentionally or not, help to conserve or maintain biodiversity while contributing to vital, resilient communities (including for instance traditional land tenure systems, resource use and management practices, ritual and ceremonial practices, agricultural practices, etc.).
2. *Indigenous, traditional, or local ecological knowledge*: Focus on ecological knowledge as related to biodiversity conservation and to the sustainability of local communities.
3. *Maintenance or revitalization of indigenous or local languages*: Focus on local language(s) as relevant to biodiversity conservation and community vitality and resilience, through communication and transmission of traditional ecological knowledge and practices.
4. *Biocultural diversity policy*: Focus on affecting existing policies or developing new policies related to biocultural diversity conservation at local, national or international levels.

As we were aware that in some cases the distinction among these criteria and areas of emphasis might be somewhat artificial, in distributing the survey we stressed that these categories were mostly a device to identify where a project placed its main focus, and that there might be overlaps with other categories. We also wanted to remain open to the possibility that new and different factors and dimensions would emerge in the course of the work, so that our working criteria and classifications would have to be modified and/or expanded. We intended our analytical framework to remain flexible and receptive to input and feedback from survey participants, as well as from reviewers and other commentators.

The survey

Phase I of the survey started in early 2004 with a first round of dissemination of the survey materials (see Appendix 2). This initial round yielded 31 suitable projects; 14 more projects were added after a second round of dissemination later that year, bringing the total of biocultural projects included here to 45. The 45 projects cover all continents, with only Europe being under-represented. In spite of specific dissemination efforts, particularly in the Nordic countries and Russia, we received only one project for Europe, with a focus on Mediterranean antiquity. This circumstance is purely an artefact of the survey process, and in no way implies that at present Europe is devoid of indigenous and traditional cultures. This significant gap notwithstanding, we consider our sample of projects, programmes and initiatives with a focus on biocultural diversity conservation to be representative, although by no means exhaustive. Indeed, as we mentioned in the Introduction, this is certainly but the 'tip of the iceberg' of the integrative biocultural work that is taking place worldwide.

After receiving the initial materials on these 45 projects, we inventoried and classified them and identified aspects on which we wished to obtain additional information from the contributors. Phase II of the survey consisted of more in-depth exchanges with respondents, in order to clarify some of the information provided by them, add more questions, and request feedback. In particular, we solicited input on two critical issues: (1) the level of community participation in the project, and (2) whether and how the project sought to establish or support methods or institutions for the intergenerational transmission and maintenance of traditional knowledge and values.

Another set of questions that we thought relevant for more in-depth analysis was later distributed to survey participants in table form (see Appendix 2). These questions touched on issues such as indicators of loss of biological and cultural diversity; indicators of persistence and resilience of cultures; methods of transmission of languages and cultural knowledge, beliefs and practices; and connections to sustainable livelihoods. We also queried survey respondents about what they thought was most crucial to share about their respective projects in terms of challenges, successes and lessons learned in the application of an integrated biocultural paradigm. In particular, we wished to hear whether survey respondents found that taking a bioculturally oriented approach to biodiversity conservation contributes tangibly to affirming cultural knowledge, practices and languages; and vice versa, whether supporting cultural knowledge, practices and languages assists biodiversity conservation objectives. We also asked contributors for stories in some of the project participants' own voices, as well as for project photos and other visuals.

Through this correspondence, we acquired a better understanding of the projects, the people involved in them, and their motivations and goals in participating in the survey. This extensive interaction with and data gathering from survey respondents forms the basis for the description of the individual projects found in Chapter 4, the cross-cutting analysis in Chapter 5, the examination of lessons learned in Chapter 6, and the identification of gaps and future directions in Chapter 7.

Analysis of results

Preliminary analysis of the results began as soon as we had enough information on the first round of projects. With the consent of all contributors, we circulated a draft report among them, asking them to provide feedback and comments on it. The draft also included the contacts for each project, so that each contributor could become aware of the other contributors and communicate with one another if they so wished. After this internal review process, we posted the draft report on Terralingua's website, where visitors to our site also viewed it. Some of these visitors became new contributors in the second round of the project. The online draft report was kept updated with the new contributions. The final project descriptions were circulated to the respective contributors for their approval.

4

Overview of the Projects

Compiled by Ellen Woodley

In this chapter, we present a description of each of the 45 projects yielded by our survey. The descriptions are based on the survey materials and subsequent information received from our collaborators. The text of each description is a partially modified and edited version of the written materials provided by the collaborators, in some cases with the addition of materials culled from project documents and websites. While each description contains the basic information about the projects that was requested in the survey, some of the contributors were subsequently able to send more extensive materials. Although the more detailed descriptions allow the reader to get a deeper appreciation of the context, challenges and opportunities of the related projects, we have included all information received, as each project is informative and valuable in its own right.

The project descriptions below are ordered by world region, identified according to the practice of the UN Permanent Forum on Indigenous Issues and the Indigenous Global Caucus. These sources list the following regions as representative world geographic areas: Africa; Arctic (including Alaska, Northern Canada, Greenland, Northern Scandinavia and Northern Russia); Asia; Europe; Latin America and the Caribbean; North America; Pacific. As none of the projects surveyed here is located in the Caribbean, below we refer only to 'Latin America' instead of 'Latin America and the Caribbean'. A few of the projects surveyed are global in scope, and are included here under the label 'Global'. The number in brackets after each project's name is a unique identifier that cross-links each project to its respective listings in the two synoptic tables in Appendix 1. The numbers also cross-reference the projects to their respective locations on a world map (see Plate 4).

Africa

Taking Conservation into Our Own Hands: Forest Protection and Management by Highland Communities in Cameroon

Project contributors: Jonathan Barnard and John DeMarco

Figure 4.1 *The forest–farm boundary at the Kilum-Ijim forest was agreed by participatory decision making with local communities*

Credit: Jonathan Barnard / BirdLife International

Cameroon is among the top ten countries in Africa for high biodiversity and cultural diversity. The rich montane forests of the Cameroon Mountains have high numbers of endemic plant, bird, amphibian, reptile, mammal and insect species. However, some areas have been cleared, such as the Bamenda Highlands montane forest, where very little remains due to years of logging, farming and grazing. It is estimated that if clearing had continued unabated, the Kilum-Ijim Forest (the largest remaining patch of Bamenda Highlands montane forest) might have completely disappeared by 1997. The Kilum-Ijim Forest Project is considered to be one of the pioneers of community forestry in Cameroon and is widely regarded as a model of how communities can manage their forests for both biodiversity conservation and to meet their own needs. When other communities learned of what was happening at Kilum-Ijim, they came to visit the project in order to learn more about conserving their own forests.

The 'Bamenda Highlands Forest Project' (4) was set up as a collaborative partnership between BirdLife International, the government of Cameroon and the communities of Cameroon's Bamenda Highlands, to help those interested communities outside Kilum-Ijim. Rather than imposing conservation on the communities involved, the project's approach was to encourage people to share the many reasons that they already had for valuing the forest, including reasons that were not widely known or were nearly forgotten by the communities themselves. Now there are more than 20 forest management institutions (FMIs) along with traditional management institutions that are active in the region, directly managing the forests without project assistance. The FMIs are community-based organizations (CBOs) that are necessary for the Community Forestry law in Cameroon. They perform a number of roles relating to the planning and management of the forest and reporting to the Government, and people are elected from the local communities concerned to fulfil these roles. The project facilitated the establishment of these CBOs, and then helped them in their development, ensuring due regard to governance and transparency. The FMIs are still functioning without the presence of the BirdLife project.

The largest tribal groups in the area decided to form two umbrella organizations to support each other better: the Association of Kom Forest Management Institutions and the Association of Oku Forest Management Institutions. There are also several local non-govermental organizations (NGOs) now active in supporting communities for local forest management, most of which were initially assisted by the project in terms of capacity building. Local practices, beliefs and languages associated with biodiversity have been revitalized, and forest boundaries and biodiversity have stabilized. While not an explicit objective of the project, the frequent and ongoing discussions about the forest have helped to revive and pass on local knowledge and elements of language to more people, including younger generations. All written materials and all species names are in the local language. Special attention is given to key medicinal plants from the forest, whose loss would have a major impact on the local practice of traditional medicine. Special efforts are made to protect an endemic local bird of great cultural significance, Bannerman's Turaco, which symbolizes the close links between culture and biodiversity, and whose extinction would have a significant impact on local practices. Under its obligation as signatory to the Convention on Biological Diversity (CBD), the national government has supported the revitalization of the important cultural values of these forests, which would have probably been lost to extractive forestry practices.

Indigenous Sacred Sites and Biocultural Diversity: A Case Study from Southwestern Ethiopia

Project Contributor: Desalegn Desissa

Figure 4.2 *Community gathering in the Dorbo sacred pasture land to get blessing from indigenous religious leaders (sitting in the front row)*

Credit: Desalegn Desissa

Sacred lands in southwestern Ethiopia are in distress, due to the lack of respect for indigenous spirituality and the failure of local government bodies to protect its indigenous peoples and their religious practices, as well as owing to pressures from tree cutting, cattle grazing and other forest encroachments. In response to these threats, a cultural movement is emerging at the grassroots level and among academic institutions and non-governmental organizations whose focus is to recapture 'whole indigenous landscapes' and their belief systems (see Plate 9).

The indigenous Gamo peoples of Ethiopia have a long history of close association with nature, and their practices of worshipping nature continue today through the veneration of sacred sites (sacred natural forests, burial grounds, ponds, streams and other landscape features), which are the link between nature, culture and spiritual realms. Traditional religion is based on a system of taboos concerning the spirits that are believed to control the sacred sites. These traditional spiritual values have served to prevent people from over-exploiting certain areas. However, these customs and values are now changing because of the abandonment of traditional beliefs and the adoption of monotheism. The expansion of monotheistic religion and the appropriation of the venue of indigenous religion are worsening. The Ethiopian constitution grants the right of worship in any religion, but in practice this is not happening at the local level.

The project 'A Collaborative Social and Biological Study with Gamo Elders of the Importance for Biocultural Diversity of Living Indigenous Sacred Sites in the Gamo Montagnard Region of Southwest Ethiopia' (43), undertaken by the Ethiopian Wildlife and Natural History Society in collaboration with Gamo elders, was designed to determine how indigenous sacred sites in the Gamo highlands maintain local biocultural diversity. Convincing the local government officials was the most challenging part of implementing the project. Most of them practise monotheistic religion and resisted the request of the research team to work with them on sacred site protection. However, after a series of discussions with concerned government officials, the work got underway.

To achieve the objective of minimizing the pressures on sacred sites and traditional beliefs, the project team has undertaken exhaustive field research and awareness raising. The team has categorized and mapped sacred sites that are still managed by the traditional custodians. In the first phase of the project, nearly 645 sacred sites were identified, described and mapped. Of these, 272 are sacred forests ranging from 0.5ha to 25ha, where over 792 plant species belonging to 149 families have been identified, including 19 endemic species and 4 species that are otherwise absent or rare in the rest of the region. The focus on the conservation potential of the traditional belief system is one way to convince both national governments and local communities of the value of local traditions.

The second phase of the project is a practical extension based on the findings from the first phase. To this effect, an organization, Friends of Gamo Gofa Sacred Sites Association, was established to give legal backing to the custodians of the sacred sites. The association consists of Gamo and Gofa indigenous intellectuals and aims to help custodians protect their sacred sites. The establishment of individual nursery sites in some communities to help restore degraded sacred forests has been successful. Awareness raising has been successful on many fronts, including seminars given to students and university staff on the importance of the culture and biodiversity of sacred sites, and workshop presentations given to decision makers on the importance of sacred sites for culture and biodiversity conservation. The workshop with decision makers allowed for networking and idea sharing among formal and grassroots opinion leaders, and for increased biocultural diversity awareness among decision makers. Another major success has been support for people to undertake ritual festivals. People have been gratified that their indigenous religion is coming out into the open after 30 years of suppression. The most touching comment in response to the festival came from a community elder: 'Thank you for helping us to get back one of the most important parts of our culture which is our life. In our age the younger people do not respect ritual places and ritual materials, as a result the wrath of our ancestors came to our land which prevented us from getting good crops, milking cows and keep our children healthy.'

The elders and other community members have made a vow to protect their ritual places. Project success depends on the participation of ritual leaders, youth and community at large, so the project has been successful thus far. Since the end of 2007, the project has expanded further into the Gofa highlands, where the identification of sacred sites and mapping is currently underway. The project aims to expand to more remote areas with more marginalized people.

Biodiversity Conservation through Traditional Practices in Southwestern Ethiopia, a Hotspot of Biocultural Diversity

Project Contributor: Zerihun Woldu

The southern Rift Valley in Southwestern Ethiopia is known as one of the hotspots of biocultural diversity and of indigenous knowledge associated with the use and conservation of biodiversity through home gardens, agroforestry practices and sacred forests. The project 'Ethnobotany of Indigenous People of the Southern Rift Valley and Southwestern Ethiopia' (45) was undertaken in collaboration with the Hamar, Konso, Dassanetch, Mursi, Me'en and Dizi Indigenous Peoples of Southern Ethiopia, with support from The Christensen Fund.

The first phase of the project focused on the ethnobotanical knowledge of the Konso and Hamar peoples. Project results show that the Konso tradition of growing multipurpose indigenous trees in their crop fields and home gardens acts to maintain these trees, even after the species become rare or absent in natural stands. Such traditions promote biodiversity conservation through uses that are essential to the Konso's livelihoods. The polycultural farming system, which minimizes the risk of crop failure, is also a means of diversifying crop niches. The subtle and active processes involved in the cultivation and gradual domestication of selected useful wild plants add yet another dimension to the local agrobiodiversity. Women play an active role in maintaining agrobiodiversity, a role not commonly recognized in research and development initiatives. There is also a tradition of recognizing and ensuring the continued existence of sacred forests in the Konso area, showing that traditional leaders and traditional institutions such as religious beliefs play a vital role in conserving these natural forests. In the highly degraded Konso landscape, remnant patches of natural forests are still found because of these traditional practices. In these sacred forests, there is less deforestation, since traditional spiritual values have influenced people's behaviour and have played a role in protecting them and ensuring that some of the culturally valued trees and other medicinal plants are found on a sustained basis. Although they occupy a relatively small area, the sacred forests in Konso have greater woody species richness and taxonomic diversity than the communal grazing lands, bushlands and scrublands protected by the community. In the Hamar area, there is also a tradition of protecting large riverine trees through a system of taboos.

An important feature of the project is that it was conducted with the active participation of the indigenous peoples of the concerned communities as equal partners of the project team, and all findings, publications and patents will belong to all team partners. The project is working to introduce mechanisms of horizontal exchange of knowledge, experience and resources of useful values, knowledge and skills with the neighbouring communities.

Although the project was initiated by academic staff of the Department of Biology at Addis Ababa University, the project is determining methods for best practices of working together with indigenous peoples based on mutual trust and equal participation for the fair and equitable sharing of benefits accrued, in line with the principles espoused in the CBD.

Countering Local Knowledge Loss and Landrace Extinction in Kenya: The Case of the Bottle Gourd (*Lagenaria siceraria*)

Project Contributor: Yasuyuki Morimoto

Figure 4.3 *Women displaying* kitete *gourds and* kitete *seed necklaces at a community festival*

Credit: Yasuyuki Morimoto/Bioversity International

For the Kamba people in the Kitui District of Kenya, the bottle gourd (*Lagenaria siceraria*) and its estimated 50 landraces are part of a rich cultural history, having been cultivated for approximately 10,000 years. Known locally as *kitete*, this plant is central to the material culture of the region and has much symbolic and cultural value, as illustrated by the complex belief system that underpins the role of this species in Kamba culture (see Box 5.5 in Chapter 5). Diverse utilization was a driving force for the cultivation of so many landraces, with a total of 61 different major uses documented so far by the project 'Community-Based Documentation of Indigenous Knowledge, Awareness and Conservation of Cultural and Genetic Diversity of Bottle Gourd (*Lagenaria siceraria*) in Kitui District in Kenya' (37). Some of the uses include: *kitete* as food – some landraces are edible, typically eaten in sauces, or boiled or fried; and *kitete* as calabashes – the hollowed-out shells have traditionally been used as containers to hold water, honey, milk and perfume, to name but a few items. The shells have also been used for many other purposes: beehives, washbasins, animal traps, musical instruments and masks. The beautifully decorated bottle gourd is also a popular souvenir and is sold in tourist markets in cities such as Nairobi (see Plate 10).

Recently, however, these multiple uses and the value of *kitete* have been greatly undermined by the use of plastic containers. This is resulting in the erosion of local knowledge and cultural practices surrounding this species, leading to it being threatened with extinction. The Kamba culture is intricately intertwined with the *kitete* landraces, and therefore loss of the knowledge of *kitete* threatens the associated local culture, customs and identity and will have a far-reaching impact on the community.

In 2001, the Kyanika Adult Women's Group (KAWG), a local women's group, in partnership with Bioversity International (IPGRI) and the National Museums of Kenya, initiated a two-year project aimed at conserving *kitete* diversity and culture. Other objectives are to generate additional income by promoting uses of *kitete*, consolidating access to *kitete* landraces and retaining the indigenous knowledge of *kitete* within the local communities. During the project, nearly 200 gourd landraces were collected and taken for cataloguing and for propagation in community fields, to produce seed for distribution and exchange. The project teams also gathered information through interviews, and songs and stories were recorded on cassette and documented in a national database in the community's own language. *Kitete* landraces are also described in the local language, using approximately 70 different names. The group established a *kitete* community museum within the village, which displays various types of gourd landraces. The museum also serves as a centre that distributes and stores seeds, and acts as an education centre to provide information for school children and other visitors. In addition, the group has shared information and their experience with other groups in the district through seed fairs, knowledge competitions and joint planting activities, as well as internationally by attending workshops and symposiums. Farmers are provided with the means to document their knowledge on a specific topic on audio tape or other media, in their own language, which can then be used in scientific journals or in a national database. This approach is meant to empower the knowledge holders and to recognize their contribution concerning the validity of traditional knowledge systems at the national and scientific level, while ensuring that knowledge holders' rights are recognized.

Some challenges encountered during the project concern the sharing of biodiversity-related knowledge. Sharing knowledge is possible only as long as the people are comfortable with making that information public. When knowledge is specialized within the community and a select group of knowledge holders claim monopoly over or sole rights to the knowledge, information may be guarded. Knowledge may also be withheld when there is economic value at stake, as is the case with some medicinal plants. Another challenge was the management of the documented information, which required advanced editing and archiving skills that were lacking within the group.

The benefits of the project have been wide-ranging. The KAWG women's group now sells seeds, fruits, products such as decorated fruits, necklaces, bowls and other containers and T-shirts, which have significantly increased local income. Marketing bottle gourd products for cash is seen as an incentive to maintain and keep the crop and its diversity. Products are being ordered from local and international entrepreneurs. Incorporating bottle gourd activities in cultural events such as community festivals also helps maintain the crop diversity and related knowledge. The community's motivation for safeguarding the diversity of gourd landraces has increased and most group members now grow edible gourds, improving nutrition. The project has improved harmony among members and facilitated many neighbouring communities who wanted to form their own groups.

In 2004, the government noted the success of the project and awarded KAWG a small piece of land to establish a new community centre and shop as well as a trophy for the best community-based income-generating project in the country. Despite the fact that the project concept – the conservation of traditional crop diversity for community development – has not yet been widely recognized within local and national government policies, the project's activities are becoming better known. The project has been picked up several times by the local newspapers and awareness of the issues is spreading to other areas and countries. Other communities and countries are now applying the method and approach used for *kitete* to different crops.

The chairperson of the women's group, Mrs Jemima Kimoni, stated: 'The experience and exposure the community went through is probably the most important thing that happened to the group members and probably the longest lasting motivation in their individual minds.' She added that it has also 'helped to empower individuals and build stronger links as well as created awareness within the group and community'.

Talking the Walk in Tanzania: Language as the Missing Ingredient of Biodiversity Conservation?

Project Contributor: Samantha Ross

Figure 4.4 *A women's focus group discussing changes in plant abundance in Goka, Tanzania*

Credit: Samantha Ross

The Eastern Arc Mountain Chain in Tanzania is one of the 33 global biodiversity hotspots and provides an ideal opportunity to study biological and linguistic diversity. The range spreads from Southern Kenya to Southern Tanzania and was formed as the Rift Valley took shape creating isolated mountainous blocks replete with unique ecosystems and biodiversity, prompting the moniker 'The Galapagos of Africa'. The mountains are home to 200 endemic species of fauna and more than 800 endemic floral species, including the popular African violet (*Saintpaulia*) and Busy Lizzies (*Impatiens*), with new species still being discovered. Tanzania is also linguistically diverse, with more than 127 indigenous languages, although Kiswahili is the *lingua franca*, spoken by 95 per cent of the population. President Julius Nyerere chose Kiswahili as the national language to promote peace, unity, national identity and tribal cohesion after Independence in 1961, as it is a neutral language, not favouring one ethnic group or region over any other. The many vernacular languages are used within ethnically homogeneous groups, predominantly in family settings in rural areas.

In Tanzania, both the unique linguistic/cultural diversity and biodiversity are under threat. A major challenge concerning the safeguarding of linguistic diversity is the lack of documentation on languages and language speakers, and national linguistic policies that neglect the importance of African languages for development. Kiswahili has the advantage of being neutral, but, without support for the other languages, it dominates all walks of life – business, education, religion, entertainment and administrative duties. The local languages are not recognized in any official capacity and are actively banned from being used in education or the media. English is an additional threat, since it is the language of global development and cooperation. The views of local people on these processes of modernization and change and how these affect the younger generations reflect current feelings and can offer insights into the future of local languages and culture in the area: 'Our children don't want to learn about the plants and the environment because they watch TV and go to school. They don't have time. They want to get jobs in the big towns.' 'Religion stops our young people from learning about their traditional knowledge. They listen to that God and not ours.' 'Traditional languages are out of date.'

Tanzania's unique biodiversity is also endangered. Changes in land use to accommodate the food needs of a growing population are the cause of habitat loss. In addition, the all-pervasive reach of globalization and Westernization and the accompanying acculturation are increasing challenges for the Tanzanian population to manage their resources. Residents of the Lushoto District of Tanzania comment: 'The forest used to come right to the edge of our village. We could get everything we needed there. Now the village has got bigger. There are more people. The forest has moved away, it is smaller. More people are cutting down trees to build houses and farms. The trees are less so the rain is less too and the soil is bad. It is a problem for our farms as we can't grow enough food.' 'It is now more difficult to find plants for medicine. I have to walk further and further. It takes me much longer to gather plants than it used to.' 'We used to easily find mushrooms and plants for vegetables. They were everywhere. Now it is difficult.'

The project 'Talking the Walk: Language as the Missing Ingredient of Biodiversity Conservation? An Investigation of Plant Knowledge in the West Usambara Mountains, Tanzania' (44) was a doctoral project based at the University of East Anglia, Norwich, UK, in collaboration with the Faculty of Language and Linguistics at the University of Dar es Salaam, the Tanzanian Forest Research Institute and the Friends of Usambara cultural and ecotourism group. Research focused on Lushoto District in the West Usambara Mountains of Northeast Tanzania, an area little researched but highly threatened in terms of an increasing population living on steep-sided hills that are intensively farmed for both subsistence and cash crops. The intensity of these activities, coupled with forest encroachment for medicinal plants, supplementary foodstuffs and timber are all contributing to habitat loss, soil erosion, water shortages and the invasion of exotic species. The local population has extensive knowledge of wild plants, using them primarily as supplementary foods such as greens, mushrooms and fruit and medicinal plants.

The aim of the project was to examine the biocultural dynamics of language in Tanzania, exploring the possible links between language and indigenous education for environmental and cultural sustainability. The project focused on the role of local languages (in this case Kisambaa and Kimbugu) and traditional knowledge in conserving the local environment and contributing to livelihood resilience and economic opportunity. The project investigated if language shift is taking place as Kiswahili becomes increasingly important in daily communication and socio-economic interactions, and what the implications are of this language shift for local languages and biodiversity conservation. These findings were then related to education and biodiversity conservation policy in Tanzania.

Research results show that language shift is indeed occurring in Lushoto District. Kiswahili and English are becoming the languages of choice above the local languages Kisambaa and Kimbugu. Within traditional spheres where the local languages would customarily be used, such as around the local area, the market, and within peer groups, Kiswahili is pushing out the vernaculars into smaller and smaller arenas such as the home and among elders. This is reflected in comments from participants: 'I teach my children Kiswahili because it is the acceptable language. It is the language we speak everywhere. Everyone understands it.' 'I speak [local languages] when I meet those who know them, but Kiswahili is everywhere.' 'I speak those languages [mother tongues] at home because at school we are not allowed to speak in Kisambaa. I speak to my friends in Kiswahili.'

However, there is one domain in which the local languages are remaining dominant: the area of ethnobotanical knowledge. Discussing plants – their uses and other pertinent knowledge, such as the plants' ecological needs and locations – is better performed in the mother tongue. The inter-ethnic dominant language of Kisambaa appears to be the language most applicable for locally specific ethnobotanical knowledge and is popularly chosen for identifying plants and describing plant practices. Kisambaa is also chosen for the purpose of transferring this indigenous plant knowledge across generations.

These findings point to the need for integrated intercultural and multilingual conservation practices. Local languages are essential for transferring locally specific indigenous knowledge that is vital for conserving the local environment and for providing the local population with livelihood resilience and economic opportunities. The findings raise questions and offer insights into Tanzanian and international debates on the use of mother tongue as the language of instruction in decentralized education systems. They also shed light on the language and knowledge base best suited for use by institutions involved in biodiversity conservation, so as to put into place the most successful practices. The fieldwork section of the project was completed in 2007 and the results are currently being finalized.

Academically, the project has contributed to the field of biocultural diversity, widening its research base to include Africa. Most importantly for the local population, a series of three books will be published so that local knowledge in a format understandable to all is documented. In the process, the books will aid environmental and cultural conservation and contribute to indigenous language preservation and maintenance. The books contain local folk stories written in the two local languages alongside Kiswahili, with pictures by local artists. A third book will have photographs of locally specific economically, socially and culturally important plants, their names in the local languages, their uses, ecological niches and conservation status. A draft copy of one of the books was shown to the research participants. Their comments show that it was well received: 'I have never seen my language written down before. This is very special.' 'This is so important for us. Everyone will want one of these books to show their children and grandchildren. You must print many.' 'The pictures of these plants will help us, and our children, to take care of our environment.' 'Our language will now be known and remembered by so many people.' The books will be made available in the local area at a nominal cost. It is hoped that the popularity and success of these books will encourage other people in other areas to document their local knowledge in their own languages for future generations, inspiring and motivating the government and other funding bodies to set aside resources for similar vitally important projects.

Promoting Traditional Medicine, Indigenous Cultural Research, and African Spirituality in Uganda

Project Contributor: Sekagya Yahaya Hill

Figure 4.5 *Traditional healing centre, Uganda*

Credit: THETA News

The traditional African culture that acted as a social security system for the weaker sectors of society has greatly eroded. In Uganda, the use of herbal medicine was labelled as 'backward, uncivilized and unholy' during the colonial era, and traditional healers suffered much humiliation. However, the knowledge of herbal medicine did not diminish entirely, but rather flourished underground (Tumanyire, 2002). In the project 'Promotion of Traditional Medicine and Indigenous Cultural Research and African Spirituality' (19), PROMETRA Uganda, a Ugandan NGO, works to protect and nurture the medicinal plants that are important to traditional healers according to traditional spiritual concepts, beliefs and practices. This ensures conservation and the sustainable use of biodiversity by local people, specifically healers who use these plants. The project also encourages the documentation and recording of traditional information in the local languages.

The traditional healers group, comprised of about 100 healers from different areas, meets in a large forested area, a site that is becoming a traditional healing and cultural demonstration institute. It is in this culturally diverse setting that the healers share and discuss the dependence on the local environment for food, medicine and all aspects of life as well as the sustainable use of their environment and their cultural approach to biodiversity conservation. This approach is based on traditional spiritual concepts, beliefs and practices and on understanding the meaning and purpose of life. In this working relationship, where knowledge is shared and collective memory developed, traditional healing methods are improved upon and local biodiversity is conserved. Traditional health practitioners know the rhythm that ensures, as project contributors stated, 'the human–nature–cosmos balance, in its symbolic and transcendent relationship with the sacred'. As one Ugandan visitor to the project commented: '[The project is a] positive direction towards culture, knowledge and plant diversity protection.'

Wild Resources and Cultural Values: Implications for Biocultural Diversity in South Africa

Project Contributor: Michelle Cocks

Figure 4.6 *The woodpile is the Xhosa women's status symbol and cultural totem*

Credit: Tony Dold

Since the 1980s, the government of South Africa has taken a more people-centred approach to conservation, and most legislation has been updated to articulate the need for the participation of local people in the management of biodiversity both within communal areas and on state-owned land (Kepe, 1999; Campbell and Shackleton, 2001). Despite the recognition in South Africa that culture is intricately bound to the use and management of biodiversity (Bernard, 2003; Fabricius and Koch, 2004), however, the use of culture as a tool in conservation strategies has not as yet been explored within the South African context (Cocks, 2006).

South Africa offers an excellent opportunity to observe whether and to what extent the effects of cultural values on biodiversity are preserved under non-traditional conditions, as the country witnessed 46 years of turbulent political history, during which time the state forcibly moved more than 3.5 million people into 'homelands', established under the apartheid regime. Consequently, in this context the concept of 'local communities' seldom represents people who have historical continuity with pre-colonial societies. In contrast, they are completely reliant on the national economy. In response, the project 'The Significance of Non-Timber Forest Product Utilization and Cultural Practices in Rural and Urban Households: Implications for Biocultural Diversity' (25) aimed to assess the importance of biodiversity with respect to cultural and utilitarian values among different categories of non-traditional communities in South Africa and to evaluate factors that contribute to the persistent use of biodiversity for cultural practices. All of the wild resources used in the study area were identified with their vernacular names. The project also aimed to reflect on how the cultural values of biodiversity could contribute towards biodiversity conservation. The project was conducted from 2000 to 2005 in South Africa for a doctoral research based at Wageningen University in The Netherlands.

The study demonstrated that the use and value attached to natural-resource-based goods remains significant despite increasing urbanization in the study area. In urban areas, 96 plant species are used regularly, and 85 per cent of these are used for cultural purposes. The importance of natural resources in fulfilling household members' cultural needs was reiterated by the finding that even wealthy households in both the rural and urban communities continue to utilize natural resources for cultural purposes. This indicates that the use of natural resources for cultural purposes transcends both economic status and the rural–urban divide.

An example of natural resource use for cultural purposes is the construction and maintenance of cultural artefacts such as *kraals*. In the *kraals* or 'temples', ritual sacrifices are performed, which represent the most important and effective form of communion with the ancestral spirits. The rituals are performed to elicit ancestral blessings and protection from malevolent forces such as sorcery, and they invariably involve the slaughter of a domestic animal, usually an ox or a goat (Wilson et al, 1952; Poland et al, 2003). Over half of the urban households interviewed contributed towards the maintenance of a *kraal*, and in the rural study area almost 80 per cent of the households own and maintain a *kraal*. Of the rural households surveyed, only 47 per cent own livestock, demonstrating that *kraals* are not just a livestock enclosure for the majority of households, but instead hold cultural significance.

The fact that many of these practices are still being maintained by a significant number of urban people demonstrates that cultural values concerning the use of wild plant resources are not restricted to traditional indigenous and local communities in rural areas. Thus, one does not have to live geographically close to the natural environment for it to continue to hold spiritual, social and cultural values for its users (Cocks, 2006). It appears that the continued cultural use of wild plant material is due to the fact that this use contributes to the maintenance of the cultural identity of the formally disadvantaged black people in South Africa. The research results clearly raise the question of whether the use of natural resources contributes to identity formation and/or strengthening of cultural identity (Cocks, 2006).

Limitations to community-based natural resource management are due in part to community social heterogeneity. One way to overcome that problem is to stress the cultural values of wild plants to the community as a whole, so that appreciation and management of this resource cross-cuts different sectors of the community, from the wealthiest to the non-wealthy members. Based on the key findings of the project, attempts are being made to raise awareness around the inextricable link between cultural diversity and biodiversity among students, as the preservation of both cultural heritage and biodiversity relies on young people recognizing the importance and value of nature in its broadest sense.

Europe

Ancient Botanical Knowledge as Living Knowledge: Medicinal Plants of Antiquity

Project Contributor: Alain Touwaide

Figure 4.7 *Dr Alain Touwaide at work in the Library at Soliman the Magnificent Mosque, Istanbul, examining ancient herbals and documents from which he recovers information about the ancient therapeutic uses of plants*

Credit: Emanuela Appetiti

The 'Medicinal Plants of Antiquity' programme (34) is recovering the ancient therapeutic practices of healers recorded by physicians of Classical Antiquity and the Middle Ages, such as Hippocrates, Galen and Avicenna. This research, which is conducted at the National Library in Rome, Italy, is documenting and reviving part of the heritage of humankind: the knowledge of medicinal plants and particular adaptations that were used during past ages, in order to see how it may be used today. Knowledge of this period is all too often forgotten and, in contemporary society, even the awareness of this knowledge is disappearing. A related project involves texts and plant representations from 15th- and 16th-century printed herbals that are collected at the National Library of Rome and further analysed at the Smithsonian Institution in Washington, DC, where they are added to a growing database. This programme is part of a comprehensive study on the botany, ethnobotany and ethnopharmacology of the ancient Mediterranean world. Medicinal plant knowledge that was used from antiquity to the Renaissance in the Mediterranean area is now all but forgotten and threatened with extinction, but may have a function today as a basis for understanding how adaptations of plant use in the past may be applicable in the present.

In this programme, the historical uses of natural resources, which are rooted in an experience accumulated over the centuries, are brought to light through an in-depth study of the constitution, transmission and transformation of knowledge over time, from antiquity to the birth of modern science. The programme champions a new model of data analysis: instead of considering historical data as 'fossils', it proposes to consider them as living knowledge that will lead to a better understanding of human relationships with the environment. This information might even help generate a new type of relationship with the environment that draws from the wisdom of past experience.

Asia

Indigenous Knowledge, Biodiversity Conservation, and Poverty Alleviation among Ethnic Minorities in Yunnan, China

Project Contributor: Xu Jianchu

Figure 4.8 *Within one hour, Tibetan villagers in northwest Yunnan collected more than 80 local species, which have been traditionally used for generations and are classified according to their own epistemology and knowledge systems*

Credit: Xu Jianchu

The opening up and success of economic reforms in recent decades in China have produced high and sustained economic growth rates and lifted millions of people out of poverty. Concurrent political reforms have decentralized many decision-making processes and created new democratic institutions, especially in rural areas. These changes, however, have placed additional stress on natural resources and on the livelihoods of indigenous communities in politically and economically peripheral areas. Increasing public awareness of deforestation and its links to soil erosion, loss of biodiversity, floods and other forms of environmental degradation has made protection of forest ecosystems a central government priority. Conflicts have emerged between decentralization for enhancing local livelihoods, and environmental protection for the benefit of larger-scale populations.

The success of economic reform and a relaxed political environment have acted to strengthen cultural identity and generate a revival of indigenous knowledge of particular traditional spiritual practices. However, a significant consequence of the changing economic context is the loss of intergenerational transfer of this knowledge related to conservation beliefs and practices. In Southwest China, the key challenge now is how to strengthen local (both informal and formal) institutions that can support and enhance indigenous knowledge, innovations and practices of the local cultural communities in relation to environmental and socio-economic changes.

The Yunnan Initiative, which resulted from the 2000 Cultures and Biodiversity Congress (CUBIC 2000), calls attention to the uncertainties that local and indigenous peoples face in their quest to use, nurture and sustain the ecosystems in which they live and on which they depend. The Yunnan Initiative articulated the principles and strategies for cultural and ecological conservation as well as sustainable economic development applicable to places that are culturally and biologically diverse. The initiative is based on the Code of Ethics of the International Society of Ethnobiology (http://ise.arts.ubc.ca/global_coalition/ethics.php) and endorses the CBD's call for respect of cultural and spiritual values for sustainable development. CUBIC 2000 concluded that partnerships between local groups and government, NGOs, and the business sector must be based on participatory processes and intercultural dialogue and institutional development, and aim for an interaction between local knowledge and aspects of Western knowledge for an equitable and sustainable stream of benefits.

The project 'Support of Indigenous Knowledge for the Use and Conservation of Biological Diversity of Ethnic Minorities in Three Ecological Regions in Yunnan, Southwest China' (22), completed in 2006, was based in three regions of Yunnan Province in Southwest China and supported by the German international technical cooperation agency GTZ (Gesellschaft für Technische Zusammenarbeit). The project followed the principles of the Yunnan Initiative: 'We fully realize that there are intricate and close links between biodiversity and cultural diversity in Yunnan; therefore we advocate in-situ biological diversity conservation within peoples' indigenous cultural and ecological systems' (The Yunnan Initiative, 2000). The project promoted indigenous knowledge for livelihoods and aimed to strengthen local institutions to use indigenous knowledge for the conservation and sustainable use of biodiversity. It was a cooperative project between Chinese and international organizations that served as intermediaries and strategic partners in the pilot areas. It also included representatives of ethnic minorities as well as forest and nature conservation agencies. The project emphasized the interlinkages between sustainable biodiversity management and poverty alleviation in Yunnan, covering tropical rainforest, subtropical broadleaf forest and alpine ecosystems (see Plate 5).

Challenges during project implementation were due to commercialization and the market economy. These changes have deeply affected traditional knowledge related to the use of biological resources for medicine, food and shelters, land use practices and customary institutions for governing access to natural resources. They have also created a divide between the older and younger generations in indigenous communities, due to the younger people working off-farm and moving to the cities. This has caused a rapid and often coerced removal of indigenous peoples from their once close dependence upon and rights to their immediate environment for their livelihoods. Further, indigenous knowledge is not sufficiently taken into account in the design and implementation of conservation and development schemes in which the government is involved. This failure is often explained in terms of government officials and resource managers privileging scientific knowledge over local knowledge. However, it may be that the conflict is rather between local vs. 'outside' objectives. The role of indigenous knowledge is sidelined because local people's objectives are ignored. The knowledge, skills, interest and patience to regulate at the local level are absent in the state.

Despite the fact that local users have the highly developed knowledge to manage their own resources, policies and regulations are made to favour the objectives and interests of the state. By privileging these objectives, local objectives, such as subsistence and resource-based commerce – the space in which indigenous knowledge can be exercised – are limited and indigenous knowledge is marginalized. Ultimately, local people have little control over their resources when these same resources are of value to higher-level elites and the state. An emancipatory approach to local development liberates people by creating a space of local discretionary power in which people can make decisions on their own behalf. Representation is a mechanism to bring forward the needs and aspirations as well as the knowledge of local communities. In addition, local authorities must have discretionary powers over resources and decisions of significance to local people.

The main success of this project is community-driven participatory action research. The project provided methodological training and advisory services to local project staff, representatives of ethnic minorities, resource managers and other resource persons. Traditional knowledge and different land use practices were documented on collectively developed local maps. These maps support local communities in formulating strategies for better management of natural resources according to their own needs and objectives. A longer-term success of the project has been networking. To ensure that knowledge is transferred, local and regional networks are being established. The exchange of experiences among the pilot areas and other villages is facilitated through, for instance, local seed fairs, cross-farm visits and study tours. This contributes to the institutionalization of dialogue among different actors such as experts in indigenous knowledge and scientists.

Culturally Rich Agroecosystems: Maintaining Traditional Beliefs for Food Security in Nepal

Project Contributor: Laxmi Pant

Nepalese 'rice culture' has provided important options to address the needs of ecosystems and local communities together, particularly in areas that are diverse, complex and resource poor. The cultivation of diverse landraces of rice has advantages over 'improved' rice varieties, both ecologically and culturally. Despite greater economic value of improved varieties, landraces are considered to have both symbolic and adaptive values. Farmers' selection of rice varieties that have been discouraged by scientists, for example, and their distaste for imported varieties, clearly show the strength of farmers' knowledge connected to social and ecological factors. The exchange of knowledge and traditions associated with landraces has important implications for the maintenance of the link between culture and food and thus for food security.

The study 'Linking Crop Diversity with Food Traditions and Food Security in the Hills of Nepal' (17), based at the Norwegian University of Life Sciences, in Norway, focused on subsistence farmers in the hills of Nepal, who have extensive knowledge associated with crop landraces and food traditions. The study suggests that the traditional practices of using local crop varieties in festivals and life-cycle rituals help maintain agricultural biodiversity, since specific crop landraces are preferred in traditional foods consumed during major celebrations. For example, *selroti*, a ring-like bread prepared from Gurdi and Madishe landraces of rice, is essential in major festivals, such as Dashain and Tihar, and in important life-cycle rituals, such as Bartabandha and Bibaha. Bread prepared from any other variety of rice would not be as desirable and might even be regarded as religiously impure. Project influence extends to policy guidelines on tourism training centres and menu development for hotels to use and promote traditional foods.

All project information was generated in the local language, using participatory learning tools, and later translated into English. However, the project's view was that language revitalization in and of itself is not a panacea for maintaining transmission of knowledge and practices. The maintenance of traditional landraces is critically dependent on the belief system and traditional practices continuing to be a part of the socio-cultural system. This is how agricultural biodiversity conservation is possible in culturally rich agroecosystems. As the project contributors point out, 'neither of the two goals, conserving biodiversity and sustaining cultural diversity is attainable in isolation'.

Recording Traditional Knowledge of Biodiversity for the People's Biodiversity Register of India

Project Contributor: Yogesh Gokhale

India is rich in biodiversity resources and the associated traditional knowledge of the properties and uses of these resources. However, the social, political, economic, technological and cultural milieu is changing rapidly, and this is significantly affecting the way in which India's living resources are being used. Further, India is lacking in well-organized, well-substantiated, well-documented information on this knowledge. There is a steady erosion of knowledge and practices of traditional systems – knowledge and practices that still have much to offer to humanity. The challenge is how to establish a relationship of mutual respect between traditional systems and formal science and how to synthesize the knowledge and practices of these two ways of understanding. The Indian national government considers it imperative that traditional systems of information on biodiversity and associated knowledge be documented in order to protect the interests of the 'ecosystem people' of India: people who have played a vital role in conserving the country's biodiversity, in augmenting it by developing thousands of varieties of cultivated plants and domesticated animals, and in developing a vast body of knowledge about their sustainable use (Gadgil et al, 2002).

This kind of system is now under development through the National Biological Diversity Act of India (2002), which mandates that local knowledge of biodiversity be registered in a national database, called the People's Biodiversity Register (PBR). The register is filling the need for the documentation and organization of oral and traditional knowledge that people choose to disclose, in addition to local innovation, all of which often goes unrecorded. There is little ground-level understanding of the various processes involved, and the PBR is designed to generate such an information base. Local knowledge that is being registered includes utilitarian uses of biodiversity such as for food, fodder, firewood, medicines used in the Ayurveda traditional medicinal system of India, as well as knowledge of traditional conservation practices such as sacred groves and sacred water bodies. In the last case, the sacred areas that are set aside are acknowledged by the national government of India and are given recognition as heritage sites. The register also includes local peoples' perceptions of ongoing and desired patterns of biodiversity management. Other legislation, such as the system of Panchayati Raj for the decentralization of administration and ecosystem management, gives special attention to local traditions and allows for the local-level implementation of India's biocultural policy in a coordinated effort at implementation at both local and national levels.

The project 'Local Level Ecosystem Assessment in India' (33) contributed to this process by recording species' names in the local vernacular, in order to link them to scientific nomenclature and provide critical material for claims related to intellectual property rights and access and benefit sharing concerning biodiversity, as per the provisions in the CBD. A manual called 'Ecology is for the People: Methodology Manual for People's Biodiversity Register' was produced for the National Workshop

on People's Biodiversity Register held in 2006 in Chennai, India.
The PBR is expected to serve as a tool to:

- document, monitor and provide information for sustainable management of local biodiversity resources;
- promote biodiversity-friendly development in the emerging process of decentralized management of natural resources;
- establish claims of individuals and local communities over knowledge of uses of biodiversity resources, and ensure equitable benefit sharing from the use of such knowledge and resources;
- teach environmental science and biology; and
- perpetuate and promote the development of practical ecological knowledge of local communities and of traditional sciences such as Ayurveda and Unani medicine.

The intention is that a countrywide decentralized yet networked system of information will serve several important purposes. It will for the first time create a mechanism for monitoring the fate of a variety of biodiversity resources throughout the country, be it medicinal plants, landraces of crops, breeds of regional livestock or wild relatives of cultivated plants. Such information could then form the basis of a strategy for the conservation of these resources. The information system will give full and proper credit to informants and will give recognition and encouragement to 'practical ecologists' everywhere, many of whom lack formal education, yet have a wealth of knowledge about the living world and its human uses (Gadgil et al, 2002).

Endangered Languages, Endangered Knowledge: Vanishing Voices of the Great Andamanese of India

Project Contributor: Anvita Abbi

Figure 4.9 *Peje, an Andamanese, cutting a bamboo for making bows*

Credit: Anvita Abbi

The Andamanese represent the last survivors of the pre-Neolithic population of Southeast Asia. Genetic research (Thangaraj et al, 2005) indicates that the Andamanese tribes are the remnants of the first migration from Africa that took place 70,000 years ago. Of the 50 remaining Great Andamanese people who live in the Strait Island and in the city of Port Blair, in the Union Territory of the Andaman Islands of India, there are only seven terminal speakers of the Great Andamanese language, popularly known as Jero. Even these few speakers have stopped speaking the language among themselves. The present-day Great Andamanese language is a mixed variety of three to four languages once spoken on these islands. The project 'Vanishing Voices of the Great Andamanese' (40) was funded by the Hans Rausing Endangered Language Fund under the Major Documentation Project, School of Oriental and African Studies, University of London, UK. The project highlighted the need for policy to assist in the revitalization of threatened languages and cultures. Its primary objective was to obtain first-hand knowledge of the linguistic situation of the aboriginal communities, as a basis for developing an interactive trilingual dictionary (Hindi–English–Great Andamanese). Another important reason for undertaking this project was to confirm the hypothesis that the Great Andamanese seemed to be a language distinct from the rest of the tribal languages of the islands, implying that this could have been the sixth language family of India. This has now been confirmed and corroborated by geneticists.

The project gathered oral histories, pictures of the local habitat, audio and video recordings of the surviving speakers, as well as sociolinguistic sketches. These sketches highlight local beliefs and behaviours, indigenous names of the islands and their different locations, as well as indigenous knowledge pertaining to the biodiversity that once existed in the islands, which is stored in the lexicon. Recorded information includes the names of a large variety of crabs and fish, various words pertaining to different areas of seashore and deep sea, uses of different kinds of leaves for hunting and gathering activities as well as for medicinal purposes, and local ecological knowledge of impending environmental disasters. A remarkable example in this regard is the perception that the Great Andamanese people had of the approaching tsunami that hit the region in 2004 and the means they employed to save themselves from devastation.

The trilingual and triscriptal dictionary in Great Andamanese–English–Hindi has now been completed and will soon be published. Sample pages can be found on the project's website (www.andamanese.net/dictionary.htm). The dictionary includes 4100 words, accompanied by 400 colour pictures and more than 900 sound files. It is rich in detailed ethnographic information and ecological knowledge that the Andamanese still possess. When a demonstration of the dictionary was made to the Great Andamanese tribes, there was a great sense of happiness and pride among them. Some of the elders were immensely thankful to the project leaders for undertaking the work, something that could not have been accomplished without the help of the speakers themselves.

Another major outcome of this project is a comprehensive grammar of the Great Andamanese language, which is still in the process of being completed. Extensive video and audio recordings of narrations, songs, tales and dialogues in the natural surroundings of the tribes are documented. Other books include *Where have all the Speakers Gone? A Sociolinguistic Study of the Great Andamanese* (Abbi et al, 2007) and *Endangered Languages of the Andaman Islands* (Abbi, 2006) with an accompanying CD-ROM that presents, for the first time, the sounds and pictures of the tribes in their natural surroundings, serving as a rare audiovisual account.

A small picture book of photographs of the tribes in their local environments recording birth, death, cultural ceremonies and life in both rural and urban areas has also been published. A first-ever CD of Folk Songs of Great Andamanese was launched in March 2008 in Port Blair and a copy of it was distributed to every household of the tribe in a function held in the school where most of the Andamanese children study. Also, no one in the community had heard any folk tale or story in the last 50 years, as when a language dies the art of storytelling and the indigenous folk tales die along with it. However, with great difficulty, ten short stories were recorded in multiple sessions from the elders of the society. A collection of these folk tales is being published. Individual stories in the form of individual books for children are in press. Each book will be colourfully illustrated and translated into 18 official languages of India. These stories will reach a large number of people to share knowledge of one of the most ancient civilizations of the world. To impart literacy among the young tribal children, the first Book of Letters was written in the Devanagari script, the script that is officially accepted in the state of Andaman and Nicobar. The book contains multicoloured pictures and was distributed to every child in the community. The local administration has acknowledged that the project has brought results. The teaching of the language has been introduced to small children studying in the Nursery School of the Strait Island.

There were significant challenges in the task of documenting this language. First, there was a lack of will on the part of the community and the government of India, who did not come forward to facilitate the project. A host of bureaucratic hurdles had to be overcome to achieve the project's goals. Some of the local officials who were to issue passes to travel to Strait Island would not oblige despite permits from the Ministry of Tribal Welfare and the Ministry of Human Resource and Development. Once the work started and the community members understood the motive and realized the importance of the work, they were ready to assist.

Another challenge was the lifestyle of the male members of the community. Alcoholism has had a negative impact on the society, and most often male members were unavailable for interviews because of drinking. There are mixed opinions on the project: the elders are sad to see that the heritage language is virtually dead, since their children do not understand it. They are also concerned that the youth no longer know how to make a boat or hunt in the sea or in the forest. However, most of the members see no problem with losing their language, as they do not think that their language is of any importance to the modern world.

Despite the shortcomings, it can be said that the project has been very successful. It achieved much more beyond the initial goal of language documentation. The main reason for its success was two exceptionally helpful consultants, who were more than willing to assist with data elicitation. One of the male research assistants thinks that if more male members were available, more could have been accomplished. The consultants involved in the project felt that the government should take the initiative to hire them to teach the basic Great Andamanese in schools and introduce a course in the school on Andamanese culture. Nao Jr., a Great Andamanese speaker, commented: 'Our own children do not understand us. We should tell them what we are and what we speak.'

Local Knowledge and Self-Determination for Conservation: The Case of the Irular of Tamil Nadu, India

Project Contributor: C. Manjula

Irular people inhabiting the southern part of India are one of the 635 indigenous tribal communities of the country. The population of indigenous tribal peoples in India, known collectively as Adivasis (original inhabitants), is estimated to be over 84 million people. Despite these high numbers, these communities usually live on the margins of society, eking out a living by collecting subsistence materials from the forests, hunting small game and working as daily wage earners. However, alienation of people from the forests started during the British colonial period when forests came to be 'properties of the government' rather than community owned. This continued through independent India. There were some attempts at addressing the injustice meted to such communities. For example, the State Forest Department was willing to come forward and work in close collaboration with such communities through NGOs and community-based organizations to reforest degraded forestland and give back some of the income accrued to the local community.

The research study 'Plant Resources: Traditional Knowledge of Irulars of Northern Tamil Nadu' (32) was part of a doctoral programme that took place from 2000 to 2004 at the University of Madras in India, with partial support from the Conservation Foundation, UK. The project sought to document the wealth of knowledge of Irular people in six northern districts of Tamil Nadu. The study was the continuation of a project of the Irular Tribal Women's Welfare Society, an NGO that works with the Irular people on their plant knowledge and helps them in economic and social terms. The study documented the knowledge of 62 Irular healers, and showed that they use around 388 plant species for food and medicinal purposes. It also determined how local knowledge of plant biodiversity used for medicines, food, hunting and ceremonial purposes acts to conserve biodiversity. The Irular healers pray to their natural environment in order to ask for forgiveness for taking or cutting the plants. They also are careful to take only what they need: if roots are needed for medicines, for example, they are very careful about collecting only the quantity necessary for treatment. As well, they make the effort to try to plant and maintain uncommon species. This conservation ethic has changed, however, due to government departments taking over the responsibility for conservation, which was traditionally the communities' responsibility.

Two important observations made during the study point to the importance of self-determination and retaining local knowledge for conservation. Once the responsibility for resource conservation was taken away from the community and transferred to government departments, there was less incentive to conserve, even with current efforts by government departments to ensure community participation. In addition, with the loss of knowledge – the youth of the Irular community have acquired only a fraction of their parents' knowledge regarding the local flora – there is little or no interest among youth in conserving local biodiversity. The younger generation sees the traditional practices and beliefs, especially those related to healing plants, as not 'modern' like imported medicine is perceived to be. The research also focused on spiritual beliefs, the role of gender among knowledge holders, and the passing on of ethnobotanical knowledge, both traditional and current, with the aim of generating an appreciation for the use of plants and thus reviving this use among the younger generation.

Countering the Loss of Knowledge, Practices and Species on Flores Island, Indonesia

Project Contributors: Jeanine Pfeiffer with the Tado Community, the Waerebo Community and Elizabeth Gish

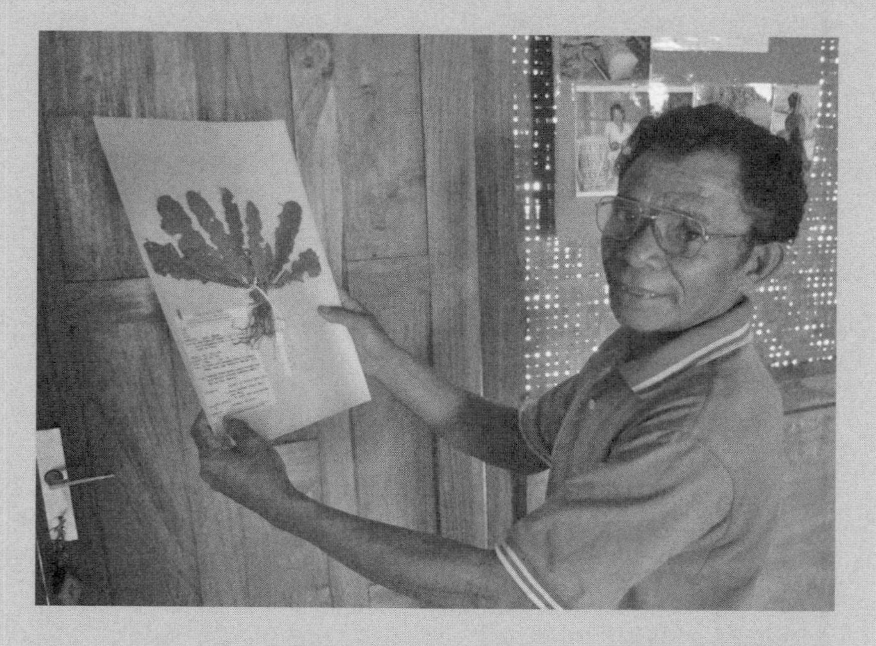

Figure 4.10 *Agustinus Angkol, a traditional Tado elder and herbarium researcher, showing one of the medicinal plants he collected*

Credit: Jeanine Pfeiffer

Tado and Waerebo are Manggarai ethnic communities located on Flores Island in East Nusa Tenggara province, eastern Indonesia. Despite being linguistically, culturally and ecologically rich, East Nusa Tenggara is perhaps the most neglected region of Indonesia. Manggarai traditional knowledge and practices are gradually being eroded due to political, economic, cultural and ecological pressures. Government support of individualized landownership certificates (versus communal lands administered by a council of tribal elders) and promotion of industrialized hybrid crop varieties have nearly wiped out the traditional circular *lingko* fields, and led to localized and regional extinction of upland heirloom rice landraces such as rain-fed rice (*mavo*). The national government also promotes non-native trees as cash crops (for example, cashew, coffee, eucalyptus) over the maintenance of native tree species.

The biggest obstacles faced in sustaining the natural world, language and culture of Manggarai are: increasing population and the need for wood for new houses which puts stress on the local forests; the increased need for food that is causing land conversion from forests to agriculture; and foreign cultures coming into the community, which bring about change in the local languages and dialects. Further, widespread conversion to Catholicism is leading to the loss of nature-based rituals related to sacred trees and stone monoliths, ignorance about the plant and animal products used in the rituals, and the denial of knowledge of ceremonial practices performed during those rituals. Knowledge of the more formalized, ceremonial (*adat*) language is in decline. As well, economic reasons are making it more difficult to perform traditional rituals involving animal sacrifices because the price of pigs and oxen increases every year.

Increasing rarity of culturally important flora and fauna that have been over-harvested or lost due to habitat destruction or invasive species is resulting in younger Manggarai generations that lack cognizance of hundreds of species their elders were intimately familiar with. Commercialization of the traditional whip dance (*caci*) for Indonesian and foreign tourists disassociates a whole suite of rituals from their ancestral cultural meanings. In the past, *caci* was only performed on auspicious dates or for deep cultural reasons approved by the elders. Other problems contributing to the loss of cultural diversity include shame or lack of interest by younger generations in 'old-style' dress, remedies or foods, as well as the introduction of plastic products resulting in a greater dependency on imported goods. As a result, the knowledge of how to make traditional meals, sing ancestral songs, recount their genealogical lineage or to prepare and administer herbal medicines is often no longer practised and is nearly lost.

The Ethnobotanical Conservation Organization for Southeast Asia (ECO-SEA, www.ecosea.org) promotes conservation, education and scientific research related to indigenous biological and cultural diversity. ECO-SEA began collaborative research with the Tado community in 1999, and with the Waerebo community in 2006. The 'Tado Cultural Ecology Conservation Program' (23) involves an on-site facility (computer lab, herbarium, insectarium and resource library) and a scientific research programme administered by local people. The Tado Community Training and Research Centre (*Pusat Penelitian dan Pendidikan Mayasarakat Tado* or P3MT) is the base for ongoing research to document native species and local knowledge about them. The programme has so far documented and revitalized 600 ethnobiological practices, involving over 200 plant species, 50 animal species and 20 insect species, and is publishing all related documents in the threatened local Kempo Manggarai language. The Tado have also mapped their ancestral lands using GPS and photo-documentation. A sub-project, the 'Tado Upland Rice Conservation Project', involves ethnographic, molecular and field research to identify and conserve traditional varieties of *mavo* grown by the Tado. Quantitative nutritional research is helping to revive traditional dishes, and qualitative anthropological research is reviving traditional stories, songs and narratives.

Over 30 Tado people and 19 Waerebo people have been trained as research associates and receive a small stipend for their work. These farmer research associates focus on Manggarai agriculture, folklore and history, traditional food and health systems, cultural ecology and the parataxonomy of plants, fungi, mammals, birds, amphibians, reptiles and insects. Known as *staf peneliti* (research staff), Tado and

Waerebo farmer research associates collectively administer research programmes in their own communities by setting up quarterly work plans, peer-reviewing one another's work, and teleconferencing with ECO-SEA once monthly.

More recently, ECO-SEA has embarked on an effort to support community-based ecotourism (CBE) initiatives and a community-to-community exchange of knowledge about institutional development. In 2006, ECO-SEA sponsored a CBE training workshop in Tado, which was attended by an interested resident of Waerebo. A two-hour trek from Tado, Waerebo is linked to Tado through inter-marriage and is the only Manggarai community still living in traditional five-storeyed, thatched-roof, circular multi-family ancestral homes called *mbaru niang*. Waerebo has more than a decade's experience with welcoming ecotourists to their village, and its residents were excited about the ethnoecological research happening in Tado. Following the workshop, Waerebo residents founded the Waerebo Ecotourism Organization (*Lembaga Parawisata Waerebo*) to build institutional capacity for existing ecotourism ventures and to initiate their own biocultural diversity research using the Tado model. Waerebo's new research initiative will benefit from Tado's significant research experience, while Tado community members can learn much from neighbours in Waerebo about how to conduct successful ecotourism activities. In 2007 the two communities signed a memorandum of understanding to mark their dedication to working together for the long term to further develop, strengthen and interlink these projects. Following the training workshop, residents of Tado designed a CBE programme and began welcoming ecotourists to their village. Ecotourism activities in Tado invite outsiders to learn about Tado indigenous knowledge and honouring of their environment and how to use its resources. Visitors participate in making a variety of crafts, such as woven fibre mats and candlenut oil lamps, preparation of traditional foods and medicines, the use of plants in jungle survival, and a selection of ancient Manggarai rituals. Their efforts were given a significant boost in 2006–2008 by a joint Swiss–Australian aid project, West Manggarai Swiss Australia Tourism Assistance (WiSATA) to support tourism development in the West Manggarai district. WiSATA featured Tado and Waerebo in their district atlas, promotional literature and on their website (www.floreskomodo.com).

Community participation and leadership at every stage in the processes of biocultural research is fundamental to ECO-SEA's approach. Collaboration between the Tado community and ECO-SEA follows the tenets of the CBD and the UN Working Group on Indigenous Populations (UN-WGIP) Principles and Guidelines for the Protection of the Heritage of Indigenous Peoples regarding the sharing of benefits and responsibilities for the conservation of biocultural diversity. The Tado and Waerebo programmes embody the principles of the International Society of Ethnobiology (ISE) Code of Ethics. ECO-SEA has brought the ISE Code of Ethics to life in the Tado and Waerebo communities. Supporting Tado and Waerebo in the development of CBE and local research programmes enables ECO-SEA and its associated academic researchers to fulfil all these ethical duties.

The resulting collaboration has had a significant impact in the communities and the lives of individuals working closely with ECO-SEA. Tado and Waerebo community members have transitioned from working as research associates on projects envisioned by visiting researchers to acting as programme administrators. This capacity-building approach secures greater continuity for long-term research, better quality and a larger quantity of data collected, and genuine community interest and engagement in the documentation and protection of their indigenous/traditional/local cultures and environments. Since community members are now able to direct and coordinate research, ECO-SEA can dedicate more resources to facilitating the newer ecotourism initiatives. Ecotourism will engage and benefit a larger portion of the community, help make the research programmes financially self-sufficient, and further boost overall conservation enthusiasm.

The ongoing, steadily increasing community involvement has reaped impressive benefits. Tado and Waerebo community members serving as farmer research associates have been inspired to enact changes in their own households, settlements and villages including: documenting long-standing land tenure disputes (using a GPS and digital camera) and being persistent about taking the case through to the highest levels of government; initiating, implementing and maintaining historical restoration and water installation projects (including budgeting, grant writing, mapping, securing in-kind and matching funding, materials procurement, voluntary group labour and transparently managing the project funds down to the last bundle of palm fibre roofing thatch or pipe fittings); seeking out and attending training courses that interest them (e.g., in avian monitoring, tourism services or household technology) in district capitals; teaching each other how to use computers and data-entry software; getting themselves elected to village councils and organizing other villagers around important issues, such as channelling government funds for much-needed health centres; building more permanent and sanitary latrines.

Finally, while bolstering and celebrating traditional Manggarai life, ECO-SEA's involvement in Tado and Waerebo is also helping to inspire changes in conservative gender roles. Priority recruiting and hiring of females and staffing practices create more opportunities and empowerment for women. Both sexes share responsibility for cooking, cleaning and childcare during staff meetings, which enables female staff and research associates to fully participate in meetings and decision-making sessions. Thus, ECO-SEA is facilitating not only conservation of native biocultural diversity, but also socio-cultural change towards greater gender equality.

A Tado community member commented about the loss of biodiversity: 'If there is no collared kingfisher (*Halcyon cloris*) or if their sounds are not heard, then the farmers will lose their signs of seasonal change. It's just like a country losing its meteorology and geophysics department.' When asked what the impact of losing their language would be, a Waerebo respondent had this to say: 'If one day nobody could speak Manggarai, all of the rituals would come to an end. There is no single traditional ritual could be performed using Indonesian. Forsaking even one ritual will cause *itang* (bad karma) to befall the whole community, so just imagine what happens if all rituals are forgotten. *Itang* is the hardest, inevitable punishment in Waerebo. The presence of *itang* makes the community realize what they should do, and they fear in their heart so that they return to the traditional rituals. It's clear that the ancestors of Waerebo always watch over the life of the Waerebo people. They will give warnings if the people go astray from their customs.'

Life with Crocodiles: Reintroducing Human–Wildlife Coexistence in the Philippines

Project Contributor: Jan van der Ploeg

The Northern Sierra Madre on the island of Luzon, Philippines, is one of the most ecologically valuable areas in the world. The area is also under severe threat from logging, destructive fishing, agricultural conversion, infrastructure development and hunting, all of which threaten biodiversity in the last forest frontier on Luzon. Rural communities depend heavily on ecosystem functions and forest products. One of the most severely threatened species in the region is the Philippine crocodile (*Crocodylus mindorensis*), now critically endangered throughout the Philippines. Over-hunting of crocodiles for the leather industry, large-scale habitat destruction (including wetland drainage and conversion to irrigated rice – a predominant crop with the 'green revolution'), and the introduction and widespread use of destructive fishing methods (dynamite, better nets, electro-fishing), all have contributed to this species' drastic decline (Weerd and Ploeg, 2004).

Another important contributing factor has been the loss of indigenous peoples' traditions and understanding of the species, including ancestral beliefs that once maintained crocodile populations. Local people, particularly fishers, traditionally were knowledgeable about the behaviour and ecology of the crocodile and its habitat (wetland ecosystems). Fishers' knowledge was generally based on opportunistic observations over a long period of time and was passed down across generations through stories and myths. Traditional beliefs and practices included strong taboos against killing and eating crocodiles. For example, in the past, the indigenous Kalinga communities in the remote area of the municipality of San Mariano in the Sierra Madre mountain range would not kill crocodiles because they believed the crocodile would take revenge through powerful spirits. People would make offerings to crocodiles in religious ceremonies or before crossing rivers, showing the veneration local communities had for crocodiles.

Over the past 50 years, however, tremendous changes have occurred in the livelihoods, education and culture of local people, as well as in their environment, leading to the loss of many of these traditions. Economic circumstances, massive immigration into the region, the expansion of the state, 'modernization' and acculturation into mainstream Filipino society – including modern education that teaches little or nothing about the local environment – have all eroded traditional forms of knowledge about biodiversity. In addition, the degradation of the local environment poses severe threats to sustaining local knowledge about biodiversity, as traditional certainties about the environment are rapidly changing. Knowledge no longer has the same meaning or function in this changing social and natural environment. Further, the behaviour of immigrants sometimes appears to belie traditional knowledge. During the logging boom in the 1970s, immigrants killed crocodiles out of fear and hunters killed them for commercial purposes. Since the local Kalinga people saw no revenge from the spirits, they began to change their worldview. In turn, the decline in crocodile populations has furthered the loss of the related traditions. Diminished familiarity with this species engenders fear of the crocodile and increases the likelihood that the animal will be killed from lack of local knowledge – a clear example of the link between cultural beliefs and practices and species conservation (see Plate 11).

The project 'Crocodile Rehabilitation, Observance and Conservation' (9) is led by the MABUWAYA Foundation, a Philippine NGO established in 2003, whose name is a combination of the two Tagalog words Mabuhay (= long live) and Buwaya (= crocodile). This project is funded by the BP Conservation Leadership Program, the Critical Ecosystem Partnership Fund and the Netherlands Committee for IUCN. It links indigenous and local governments and The international conservation movement through the development of a community-based conservation strategy in partnership with the Agta and Kalinga peoples in San Mariano. The project is assisting the indigenous communities in obtaining land rights, while seeking to conserve the small and fragmented crocodile population that remains in the area. Traditional practices that were beneficial for crocodile conservation are revived and the traditional knowledge on the behaviour and ecology of the Philippine crocodile is documented. The project thus promotes past traditional practices in a contemporary context, and enshrines these cultural traditions in law.

The project also promotes scientific research on the ecology of the crocodile. Graduate students from Isabela State University (Philippines) and Leiden University (The Netherlands) conduct fieldwork in San Mariano. The MABUWAYA project also supports a public awareness campaign, with all communication material produced in the local languages, Tagalog and Ilocano. The project aims to instil a sense of pride in the presence of the Philippine crocodile and in the related cultural traditions, thus making a crucial link between species conservation and the culture and identity of the people (Ploeg et al, 2008). The Local Government Unit (LGU) of San Mariano has become an active partner in crocodile conservation. It has declared the Philippine crocodile the flagship species of the municipality, enacted local ordinances that protect the crocodiles and established the very first Philippine crocodile sanctuary in the country, covering one of the breeding areas (Ploeg and Weerd, 2004). Three crocodile sanctuaries and eleven fish sanctuaries have now been declared and delineated. The sanctuaries are co-managed by local communities. The conservation programme is intended to be entirely community-based: without their full local consent, the LGU of San Mariano cannot declare any sanctuaries (see www.cvped.org/croc.php).

The main challenges to this ongoing project are poverty and weak governance. A small conservation project can do little to alleviate poverty among 40,000 people, who earn less than US$2/day. It is necessary to empower village councils to actively enforce environmental legislation that protects wetland resources on which the community depends. The Philippine crocodile remains critically endangered, and finding structural funding for conservation is a major issue. Livelihoods and incomes are not improving in San Mariano despite fundamental changes in the landscape: mining and biofuel plantations are new developments in the area with potential harmful effects on people, wetlands and crocodiles. Civil insurgency is another problem in the area. Maoist insurgents and the army are fighting for control of areas where the project is working and communities are often caught in the conflict.

Despite these challenges, there is some degree of success currently, marked by the number of crocodiles, which now stands at 70. There were four new crocodile nests in 2008 alone, showing good recovery in the wild. The project is optimistic in that with four more years there will be more than 100 non-hatchling crocodiles surviving in the wild in the northern Sierra Madre. The crocodile sanctuaries also strengthen socio-economic development. There is growing societal support to stop the use of destructive fishing methods (in San Mariano people fished with dynamite, electricity and pesticides). Enforcing environmental legislation helps communities to fish in a more sustainable way and crocodiles are no longer purposely killed. The Philippine crocodile has become the flagship of local environmental stewardship.

Pacific

Caring for Country: Transmission of Aboriginal Environmental Knowledge in Western Australia

Project Contributor: Kimberley Language Resource Centre Aboriginal Corporation

Figure 4.11 *Barbara Stuart and Bonnie Marisha Sampi with* Barndu *(water goanna) and* Jalij *(freshwater prawn)*

Credit: Kimberley Language Resource Centre

The Kimberley region of Western Australia is one of the most linguistically diverse areas of Australia. At least 42 languages, plus dialects, were identified post-colonization. According to 2009 data from the Kimberley Development Commission (www.kdc.wa.gov.au), Aboriginal people form almost 48 per cent of the population of the region, or roughly 16,500 people. The Department of Environment and Conservation has also acknowledged this region as an area of great biodiversity. The Kimberley Language Resource Centre (KLRC), the first regional language centre established in Australia, was incorporated in 1985. In over 24 years of operation it has cemented its status with Aboriginal people as the most representative body for

Kimberley Aboriginal languages. It services an area of 422,000km^2, including six towns and approximately 50 Remote Aboriginal Communities. It is governed by an elected board of 12 Aboriginal directors accountable to a membership representative of the approximately 30 languages still spoken, which represent about a fifth of the remaining national languages.

The KLRC is often asked to provide linguistic support to Kimberley language groups carrying out documentation of plants and animals through other bodies working in the natural resource management (NRM) field. When collaborating with language groups and other agencies on ethnobiological projects, the KLRC takes the following position:

- ensure the development of ethnobiological resources appropriate for knowledge transmission in the community – with strong language outcomes;
- provide appropriate professional development for Aboriginal people to document their own knowledge;
- provide direct support to the community to produce language resources, e.g. DVDs, bilingual books; and
- encourage language immersion at every opportunity.

A general problem with language transmission outcomes in these kinds of projects was identified by the KLRC Board and by language groups. Often, the fast-disappearing Aboriginal languages documented during field trips figure as just a list of words in publications or other resources. The knowledge found in oral language captured in audio or audiovisual recordings remains unused because of the prevailing focus on written documentation in English. The KLRC is often left to find additional funding in order to increase the language transmission outcomes – but these funds are not easy to obtain, since ethnobiological work is primarily regarded as NRM and not as language and knowledge maintenance.

One example is the Jaru Plants and Animals project, initiated by the Kimberley Land Council and the Ord-Bonaparte Program in 2004 and involving extensive field-work in 2004–2005. The Jaru language group in the Halls Creek area has been pushing strongly for materials development from this fieldwork. Sadly, several of the elderly language speakers involved have passed away since that time. The 'Jaru Ethno-biological Language Knowledge Project' (16), located in the KLRC, was established to consolidate strong language transmission outcomes from ethnobiological documentation. The KLRC was successful in sourcing funds in 2008 (Department of the Environment, Water, Heritage and the Arts, www.arts.gov.au) and is now working with Jaru speakers to identify what kind of resources they believe will best help them pass on their ethnobiological knowledge in their language.

Younger generations in the community are being encouraged to assist with audiovisual and written resources, to reduce the reliance on non-Aboriginal linguists or other outside specialists. One group of men is now working with a local Aboriginal film-maker to develop a DVD which captures knowledge about trees in the Jaru language, and links that knowledge to how these trees are used in artefact making. A group of women are interested in making resources that capture their knowledge about bush medicines, which can be used to teach children. KLRC also strongly encourages the involvement of younger generations in bush trips and other activities to increase their immersion in the language.

Bridging the (Digital) Gap: Aboriginal and Scientific Knowledge of Biodiversity in Northern Australia

Project Contributors: Helen Verran and David Turnbull

Several groups of Australian Aboriginal peoples are seeking ways to use digital technology (computers, digital cameras, sound recordings), in particular contexts, to keep their own languages and ecological knowledge systems strong. The project 'Biocultural Diversity: Elaborating Theoretical Issues for Communities and Policy Makers' (5) is one of several related projects that were conducted in 2003–2006 within the Indigenous Knowledge and Resource Management in Northern Australia programme (www.cdu.edu.au/centres/ik/ikhome.html), coordinated through the School of Australian Indigenous Knowledge Systems at Charles Darwin University. This programme aimed to support and develop indigenous databases that maintain and enhance the strength of local languages, cultures and environments in Northern Australia, by means of research on how people are creating collective memory with computers in indigenous communities in Northern Australia. The project dealt specifically with ways to assess biodiversity by drawing on Aboriginal cultural knowledge. It addressed the challenge of how to devise forms of data collection that enable different knowledge traditions (indigenous and Western scientific) to work together. The database TAMI (Text, Audio, Movies and Images, www.cdu.edu.au/centres/ik/db_TAMI.html) stores and manages data for indigenous peoples' use. TAMI is a cataloguing type of software that provides a visually based system for people to manage their own digital resources for perpetuating collective knowledge traditions. The database system adheres to the principles and practices of indigenous knowledge production, is designed to be useful for people with little or no literacy skills, and encodes no assumptions about the nature of the world or the nature of knowledge – instead, it is the user who encodes structure into the arrangements of resources and metadata. The users themselves become the designers as they bring together resources, then group and order them, and create products, such as DVDs and printed material. The project worked at the interface between academic research and engagement in policy formulation and activism for indigenous peoples' rights.

Integrating Local and Scientific Knowledge: The Wik, Wik-Way and Kugu Ethnobiology Project in Queensland, Australia

Project Contributor: Sarah Edwards

Dramatic changes to Aboriginal societies in Australia, which started with European colonization over 200 years ago and led to severe cultural erosion and the extinction of many Aboriginal languages, continue today with globalization. Environmental degradation, as a result of ranching, mining and the influx of feral animals and invasive species, is contributing to overall loss of local knowledge and biodiversity. The change from subsistence economies to one predominantly based on 'passive welfare' has also contributed to a loss of traditional knowledge, languages and practices. In Aurukun, Cape York Peninsula, Queensland, Australia, a breakdown in traditional pedagogy among Wik Aboriginals was caused by the closure of 'sacred schools' more than 30 years ago. It was at these schools that young Wik men reaching adulthood were segregated from the rest of the community and instructed in both sacred and practical aspects of 'caring for country'. The loss in continuity of traditional knowledge is summed up well by one Wik-Alkan Traditional Owner, who lamented in 2002, before passing on: 'My parents taught me the name of every tree, every plant, every fish... In twenty years this will all be forgotten. Young people today prefer to live in the busy world.' (Aurukun Ethnobiology Database Project, 2006).

The project 'Wik, Wik-Way and Kugu Ethnobiology Project' (31), based in Aurukun, is a cross-cultural, collaborative initiative between Western-trained scientists and local experts who belong to the Wik, Wik-Way and Kugu Aboriginal groups, including local rangers from Aurukun's Land and Sea Management Centre, who mediate on behalf of Aboriginal Traditional Owners. Wik are a number of closely related Aboriginal groups linked through kinship and totemic affiliations and who speak related languages or dialects (e.g. Wik-Mungkan, Wik-Alkan and Wik-Ngathan). Kugu are similarly comprised of several closely related groups, although Kugu languages are considered to fall under the Wik umbrella term. Wik-Way are considered apart from the main Wik and Kugu grouping, having traditionally been separate culturally.

The crisis in the loss of local languages that is occurring rapidly across much of the Wet Tropics World Heritage Area (WTWHA) in northern Queensland is being addressed by means of language training programmes. An Aurukun Ethnobiology Database has been developed, which integrates local Wik knowledge with scientific data, giving parity to both. The database documents Wik and Kugu names of elements of their environment as well as local plant taxonomies and traditional land management techniques (such as the use of fire) that were being lost. The database acts as an educational tool as well as a tool for use in conservation and land management. Local biocultural understanding has contributed to the development of policy at the regional level (such as in relation to the control of feral animals and weeds), as well as of a national oceans policy. The Wik, Wik-Way and Kugu Land and Sea Management Centre has a policy of following the

ISE Code of Ethics. Additional aspects of the project are the development of tools to promote intergenerational transmission of knowledge, and the identification of potential commercial opportunities for the local community of Aurukun, based on the sustainable use of wild species.

Practical challenges that had to be overcome in this project included collecting primary Wik and Kugu data about local biodiversity for the database, since much of the traditional knowledge relating to a clan estate (including plants and animals that are found there) belongs to the Traditional Owners, is often considered sacred, and thus is rarely divulged to outsiders. To overcome this difficulty, the principal scientist and data collector in this project worked closely in partnership with Traditional Owners who led the data collection process. Further, one of the community's Songmen (main traditional knowledge custodians) early on in the project 'adopted' the scientist as his own daughter, thereby giving her kinship rights and thus allowing Wik protocols to be adhered to.

It is difficult as yet to assess the actual impact of the database and how it is being used in practice, but Wik youths used it to promote their traditional knowledge in a local ecotourism initiative. A number of the original custodians of the knowledge incorporated into the database have passed away, thus making the database a valuable legacy and ensuring the knowledge is not lost altogether. Other Aboriginal communities in northern Australia have expressed an interest in developing a similar kind of database, so the Aurukun Ethnobiology Database may serve as a prototype for other areas (Aurukun Ethnobiology Database Project, 2006).

Putting Australian Aboriginal Cultural Values on the Map: The Wet Tropics World Heritage Area as a Biocultural Landscape

Project Contributor: Bruce White

The project 'Mapping Aboriginal Cultural Values in the Wet Tropics World Heritage Area' (42) was originally supported by the Aboriginal Rainforest Council Inc. (ARC), and is now supported by The Aboriginal Rainforest Advisory Committee, which comes under the Wet Tropics Management Authority, as well as Queensland Natural Resource Management Ltd. The management authority broadly represents 18 rainforest Aboriginal tribal groups on land and cultural heritage matters across the WTWHA in north Queensland. The project objectives are to overcome the rainforest Aboriginal peoples' social, economic and cultural disadvantage in the region, to assist and ensure their future cultural survival and to help coordinate their efforts to protect and manage Aboriginal cultural heritage and values in the Wet Tropics region.

A significant element of the project has been cultural mapping, which maps Aboriginal values onto the landscape by visiting their places of origin and recording Aboriginal beliefs, knowledge, heritage and practices for future collaborative management of the region as a biocultural landscape. This is a landscape where biological diversity is intricately tied to a diversity of Aboriginal knowledge, values and practices over generations. The project anticipates that Aboriginal peoples' cultural contribution to biodiversity conservation will lead to the collaborative development of innovative, creative and informed approaches to dealing with present-day problems facing environmental scientists and land managers in the WTWHA. Guidelines for equitable partnerships between Aboriginal peoples, all levels of government and the broader community to address a wide range of social, cultural, environmental and economic issues are contained in the Aboriginal Natural Resources Management Plan. The management plan takes an approach that is different from other resource management plans, in that it raises national awareness of the pivotal role that Traditional Owners play in the ecologically sustainable development of northern Australia. In so doing, it aims to increase opportunities for and involvement of indigenous peoples in local and regional resource management.

A Cultural Heritage Information Management workshop was held for Traditional Owners in the WTWHA in November 2006. The aim of the workshop was to share ideas on how cultural heritage information and traditional knowledge are being managed within and outside the Wet Tropics region. The workshop was meant to empower Traditional Owners to provide advice on the development of appropriate design for cultural heritage information management systems in the WTWHA. The project emphasizes that it is critical for Traditional Owners in the region to be part of the information management design and direction from the beginning, and the workshop provided the first of many steps in developing the most culturally appropriate information management system. The development of a cultural heritage management system takes into account multiple uses, including use as an educational tool, a data archiving tool, a tool to monitor and manage cultural sites, areas and tracks, a tool for administering Native Title rights and responsibilities, and a tool for ensuring that significant cultural heritage information is retained. The system will potentially be available to a broad range of users with different levels of expertise and will cater for use at both regional and local scales. The Kuku Nyungkul were the first group to complete the Cultural Heritage Mapping Training in 2007. They have successfully obtained an Environfund grant to help with biodiversity and cultural management in their territories.

Cultural heritage mapping in the WTWHA continues under the aegis of Terrain Cultural Resource Management (www.terrain.org.au/programs/people-a-country/heritage-mapping.html). When the project is completed, the Commonwealth Government of Australia will be approached to renominate the WTWHA under the World Heritage Convention for relisting as a biocultural landscape. The crisis occurring with the rapid loss of Aboriginal languages and associated knowledge is a part of the imperative for this kind of action by state, commonwealth and international agencies. These Aboriginal languages can be expected to form an important part of the case for renominating and relisting the area as a biocultural landscape.

Mining and Cultural Loss: Assessing and Mitigating Impacts in Papua New Guinea

Project Contributors: Martha Macintyre and Simon Foale

Figure 4.12 *Kinami Mountain on Lihir Island, PNG*

Credit: Simon Foale

Lihir Island, Papua New Guinea (PNG) is a site for gold mining by a large multinational company – Lihir Gold Limited (LGL) – which is projected to be operating for 35 years. The mining involves open pit extraction with deep-sea tailings disposal – a system that has been strongly criticized by some international environmental groups (Macintyre and Foale, 2004). The adverse environmental impact of mining in PNG has generated major social disruption, including the loss of cultural and environmental knowledge in several areas of the country where mining has taken place. The project 'Social, Environmental and Economic Sustainability in the Context of Melanesian Mining Projects' (21) is a collaborative effort between the Australian Research Council (ARC), the University of Melbourne, and the Lihir Management Company, and is being implemented in three areas in PNG. This interdisciplinary project aims to integrate social and cultural analysis with agrarian and environmental studies, focusing on the development aspirations of local people, based on their

Figure 4.13 *Lihir mine site, PNG*

Credit: Simon Foale

understandings of social and environmental impacts at various stages of mining and on issues of long-term sustainability. The research addresses problems of cultural loss in the context of mining. It documents traditional uses of the local flora and fauna such as hunting and medicinal plants, as well as access to water, forests and land, and examines the effects of certain pressures, such as mining and population expansion, on traditional knowledge and its relation to ideas of biodiversity conservation. A local educational component is included, whereby schools participate in various research projects. The results are compiled, and findings, photographs and other relevant materials are presented in posters, booklets and videos that are made available to the schools.

The project has been considered successful, but not without various setbacks along the way. The mining company lost interest in, and reduced support for, the project after it started, and this made work more difficult for some team members. Also, there is a profound tension locally between the desire for a better standard of living – which is facilitated by royalties, compensation, employment and development programmes associated with the mine – and the negative social impacts of mining, including loss of cultural and environmental knowledge and a pervasive disruption to traditional governance structures. This tension between aspirations for a more affluent life and the obvious loss of some aspects of tradition is not limited just to Lihir, but can be seen all over Melanesia.

Countering Fish Stock Depletion through Traditional Knowledge, Tenure and Use of Marine Resources in Papua New Guinea

Project Contributors: Martha Macintyre and Simon Foale

Fish stocks around Lihir Island in PNG are threatened by over-harvesting, as determined by research conducted by Australia's Commonwealth Scientific and Industrial Research Organization. There is a real need to understand current and projected use of near-shore fishery resources in the context of rapid social and economic changes driven by a large mining operation that commenced in the area in 1997 (see previous project). The project 'Traditional Ecological Knowledge Relating to Marine Environment and Fishing on Lihir' (27) is a collaboration among communities on Lihir, the University of Melbourne, and the Resource Management in Asia-Pacific Program at the Australian National University. The project focuses on how the people of Lihir understand local marine resources, marine tenure systems and methods of use, both in the past and at present. The study examines traditional fishing techniques, ideas of ownership and management of resources and restrictions on marine exploitation associated with the local belief system. The effects of more intensive fishing, which occurs because of introduced technologies and increase in population, are communicated to the people on Lihir. Low-impact exploitation strategies are encouraged in the attempt to influence local-level policy to reduce over-exploitation of fish stocks.

Those involved in the project consider it to be a resounding success. The project was carried out with proper consultation at all levels, and interviews were conducted with requisite cultural sensitivity. The main challenge is, however, that most people aspire to a better life, materially, than they have at present, and are shifting from a subsistence economy based on fishing and growing yams to a cash-based market economy. At the same time, they are worried about and dismayed by the many negative social impacts that have accompanied mining and a greater engagement with the global economy.

Integrating Customary Tenure Systems in Marine Protected Areas: A Solomon Islands Example

Project Contributor: Shankar Aswani

Figure 4.14 *Men involved in participatory mapping exercise in Roviana, Solomon Islands*

Credit: Shankar Aswani

Protected areas presently cover less than 0.5 per cent of the land- and seascapes of the Solomon Islands. In part, this is because Solomon Islands legislation lacks specific and appropriate provisions for creating protected areas, but the creation of protected areas is also complicated by patterns of land tenure. Land use is determined by holders of customary rights to the land, namely individuals within local communities. Overall, there are well over 30 marine protected/conservation areas in the Solomon Islands, managed by NGOs and community-based organizations, as well as by the government's Environment and Conservation Division. To date, more than 40 other sites have been identified and recommended as potential marine conservation areas, deemed to be of high marine biodiversity significance (Supporting Country Action on the CBD Programme of Work on Protected Areas).

The programme called the 'Western Solomons Conservation Program' (WSCP) works in tandem with the Roviana Conservation Foundation (RCF), which is a local community-based organization established with the assistance of WSCP in the Roviana and Vona Vona Lagoons, Solomon Islands. The project 'Establishing Marine Protected Areas and Spatio-temporal Refugia' (12) is located in the Western Province of the Solomon Islands, and is a collaborative effort by researchers from the US and customary landowners. Its central objective has been to create a network of Community-Based Marine Protected Areas (CBMPAs) to conserve marine and riparian habitats in various areas of the Western lagoons in the Solomon Islands.

The protected areas strategy is based on an amalgamation between customary management and modern conservation methods. More specifically, the CBMPA sites were selected through various research strategies, including:

- an ethnographic study of regional customary sea tenure to assess, among other factors, the feasibility of implementing fisheries management in the area;
- the incorporation of the visual assessments of local photo interpreters, who identified benthic habitats, resident taxa and spatio-temporal events of biological significance, into a geographic information system (GIS) database;
- the coupling of indigenous ecological knowledge with marine science to study aspects of life history characteristics of vulnerable species;
- the incorporation of fishing time-series data (1994–2004) into the GIS to examine spatial and temporal patterns of human fishing effort and yields.

The use of customary land and sea tenure systems, which are traditional structures that set the rules for resource access using customary management practices, has reinvigorated traditional authority over peoples' marine resources, and has generated innovative governance institutions, which are being articulated with customary and statutory law.

The Roviana, Vona Vona and Marovo Lagoons and adjoining coastal zones encompass a variety of critical, biodiversity-rich habitats and species in the region and include shallow coral reefs, outer coral reef drops, grass beds, freshwater swamps, river estuaries, mangrove, coastal strand vegetation and lowland rainforests. The network currently includes 26 CBMPAs, with another four established autonomously but under the auspices of this programme. In addition, the programme has contributed to awareness raising, delivered environmental education, established the institutional infrastructure to sustain the CBMPAs, and participatory development to assist in protection of other sites on other islands where resource management is more challenging. The project is also working to legalize all protected areas at the provincial and national levels and to conduct baseline marine and social science research in the areas, as well as running field training programmes for Pacific Island students. An environmental dictionary in the Roviana vernacular describes all marine and terrestrial organisms known locally. Local confidence in the programme derives partly from the fact that it is a co-management arrangement and that it includes customary authority and practices. It therefore represents an extension and revitalization of traditional tenure practices, which people can relate to and articulate in the local cultural idiom.

In consolidating the CBMPA network, the WSCP and RCF are:

- amalgamating traditional leadership, religious moral authority and governmental legal support for region-wide control and supervision of MPA sites;
- enlarging, expanding and consolidating the MPA network;
- working towards implementing the first comprehensive plan for ecoregional marine biodiversity conservation and fisheries management (related to food security) in the Solomons;
- providing technical assistance and training in MPA monitoring locally;
- contributing innovative marine and social science research concerning the ecological and social effects of MPAs managed under customary and church-based governance;
- fostering MPA environmental education at the local, national and international levels;
- evaluating the socio-ecological impacts of the 2007 tsunami and providing information useful for future plans to attenuate the impacts of natural disasters in this region;
- gazetting all MPAs and other regional coastal co-management plans;
- working to establish a comprehensive set of guidelines for implementing marine conservation initiatives in this region.

Currently, an increasing number of communities across the Western Solomons are requesting the programme's assistance in establishing both land-based and sea-based conservation programmes. In addition, one of the largest and most powerful church denominations in the region, the Couples for Christ (CFC) has asked the programme to establish a Ministry of Environment branch within the church, which would institute ecosystem-based management plans in each CFC village and would supervise the conservation and resource management activities of each participant community. This is an unusual opportunity to achieve successful ecoregional marine and terrestrial biodiversity conservation.

A number of conservation programmes in the Western Solomons have failed due to a fundamental misunderstanding of local peoples' aspirations and the socio-cultural context in which the conservation programme was being implemented. In this regard, the WSCP and RCF have succeeded in understanding the local culture (for instance, tenurial rights) and in working with local communities as equal partners to establish conservation programmes while assisting the communities in managing their resources. As Western Solomons people talk about this programme's success, more and more communities are asking for assistance in implementing their own conservation programmes. The programme leaders believe that this is a momentous opportunity to protect marine biodiversity while also supporting the traditional beliefs and cultures of the peoples of the Western Pacific.

Taboos and Conservation: Traditional Conservation Sites in the Marshall Islands

Project Contributors: Nancy Vander Velde with Jorelik Tibon

In previous times, tribal chiefs could designate an island, a section of land or reef as being *mo*, or 'taboo'. These areas were off-limits to people in general, being reserved for only certain personages and purposes. As in other countries, however, changes in biodiversity and culture have continued to increase in recent years. The Marshall Islands' biodiversity has become threatened by invasive species, urbanization, development and climate change. Caring for traditional resources has often been neglected as the society has moved into more contemporary systems of economics and governance. Over the past few years, however, some MPAs have been established in the Marshall Islands, and continue to be established on some remote atolls such as Rongalap, Ailingnae and Rongerik. The project 'Collection and Documentation of Traditional Conservation Sites' (7), based in Majuro in the Marshall Islands and supported in part by the local government, documents the traditional knowledge and beliefs linked to traditional conservation sites and other traditionally taboo areas in the Marshall Islands. The Woja Conservation Area was recently established in part of Majuro Atoll, the capital of the modern Republic of the Marshall Islands (RMI). Being the population centre, it is probably the most visible of the current Marine Protected Areas in the country. There are roadside signs that serve to raise public awareness of the concept of modern protected areas.

Jorelik Tibon, who was Project Coordinator of the Marshall Islands Biodiversity Project (and contributor to this report), resides in the land adjacent to the MPA. He expresses his point of view about the current management of MPAs as follows:

> *Not enough attention is being given to understanding the new challenges of protected areas from the perspectives of the caretakers of these resources. In the past, when the authority and the law were vested with the ruling* iroij*, or high chiefs, the people did observe sanctions and orders issued by the* iroij*. Now law and order are held by constitutional governments on the national and local level, and therefore the governments need to be part of successful management of* mo *along with the* iroij*. Since the national constitution recognizes the rights of the* alaps *[traditional landowners], they likewise need to be involved. As landowners and other people live close to the conservation sites, they need to be part of government conservation initiatives, because they are the ones using these resources. On the other hand, for community-based conservation initiatives such as the one in Woja area, they also need the help of governments for activities in relation to which the local community lacks the expertise or resources. Assistance from the Environmental Protection Authority and the Marshall Islands Marine Resources Authority for marine surveys and incorporation of modern and scientific applications and approaches is vital. Local government can be useful for monitoring and policing the*

areas. For questions that call for more scientific studies (such as the state of water quality and likelihood of marine animal survival in the environment considering pollution and climate change impacts), additional help from outside the community and perhaps the country is required.

The project aims to integrate traditional concepts of conservation into the National Biodiversity Strategy and Action Plan of the Marshall Islands.

Reconnecting with Natural and Cultural Heritage: Flora and Fauna of the Marshall Islands

Project Contributors: Nancy Vander Velde with Jorelik Tibon

In the Marshall Islands in the Pacific, as is occurring in many other areas of the world, traditional lifestyles are being replaced by urbanized ones. This transformation, compounded by the occurrence of invasive species and other non-native species, is resulting in disconnection from local biodiverse surroundings. Much of the traditional environmental knowledge is lost, along with Marshallese languages, especially among the younger generations, who no longer know the names and uses of the local flora and fauna.

Through the project 'A Review of the Birds and Plants of Bikini Atoll, Trees of the Marshall Islands and Fish of Micronesia' (1), efforts were made to preserve some aspects of the biodiversity of the Marshalls. One result was the book *Seashells and Other Molluscs of the Marshall Islands* produced through the Republic of the Marshall Islands Historic Preservation Office and the United States National Park Service. The book presented known species of gastropods and chitons, with Marshallese names, along with traditional stories and usage. This will hopefully be followed by a similar presentation on the bivalves and cephalopods. Producing such guides to the local flora and fauna, to be made widely accessible locally, is seen as a contribution to fostering language and knowledge transmission.

Efforts have also been recently made to preserve one of the most important traditional food and general-use plant species, *Pandanus tectorius*. In much of its range, this tree is only found in a wild form, but in times past the early inhabitants of the Marshall Islands developed numerous edible cultivars. So far, through the Republic of the Marshall Islands Agriculture Division and the United States Department of Agriculture, Forest Service, over 200 names have been documented for these cultivars, and ongoing efforts are being made to locate and preserve as many of these as possible. However, some of the cultivars may have gone extinct and local knowledge of and interest in the subject seem to have been lost over recent decades. Many members of the younger generation appear to only know the names of three or four cultivars.

The people of Bikini Atoll in the Marshall Islands face an even greater challenge to maintain knowledge of their ancestral home. They evacuated their atoll during the nuclear testing in the 1940s and 1950s, and until now the land remains too radioactive for permanent habitation. In October 2003, the Kili-Bikini-Ejit Local Government sponsored the project leaders to visit Bikini to document the birds and plants of the atoll. The intent is to produce scientific papers and possibly popular books on these topics.

Other projects include the preservation of the only remaining indigenous land bird in the Marshall Islands, the Micronesian Pigeon or *mule* in Marshallese (*Ducula oceanica*), particularly the subspecies *ratakensis* that is found only in the eastern chain of atolls in the Marshall Islands. The Marshall Islands Conservation Society has been overseeing this project, which has attracted the attention of international birders. So far, work has concentrated on the birds found on Majuro Atoll, but efforts are underway to expand to other atolls where the subspecies is reported still to be found – or reported to be found in the past – with hopes of protection and even reintroduction. DNA testing is being done to assess the genetic status of the existing birds.

Teaching and Learning from an Indigenous Perspective: Knowledge and Language Revitalization in Hawaii

Project Contributor: Chad Kālepa Baybayan

A consortium of Native Hawaiian schools and education professionals is using the indigenous Hawaiian language as a medium for making connections between traditional and formal scientific knowledge within a Hawaiian paradigm – one that is grounded in practices that allow people to be self-sufficient by sustaining the environments that feed and nurture them. Those environments – the sky, air, rain, rivers, streams, wetlands, shores, reefs, deep ocean, together with people – are part of an everlasting symbiotic relationship that Native Hawaiians recognize, protect and preserve because doing so sustains the generational cycle of indigenous existence. What researchers would label 'biodiversity conservation', indigenous Hawaiians would simply call the *Kumu Honua Mauli Ola*, or 'the way we live'.

The consortium has spearheaded several initiatives under the project title 'Knowledge and Language Revitalization in Hawaii' (41). The He Lani Ko Luna Community-Based Learning Centre, located on a 10-acre farm run by 'Aha Pūnana Leo (language immersion pre-school), has hands-on learning activities that focus on *'ōlelo* (language); *lawena* (social behaviour and traditional protocols); *pili 'uhanae* (spirituality); as well as *'ike ku'una*, which is traditional knowledge that makes connections to the contemporary world. The College of Hawaiian Language at the University of Hawaii, Hilo, has long offered regular classes in traditional farming, medicinal herbs and gathering of native forest products; traditional fishing and aquaculture; and song and dance through performance to celebrate and record orally the history of the Hawaiian people. At the Nāwahīokalaniȳōpuȳu immersion school, learning occurs in the Hawaiian language and within a Hawaiian paradigm. The curriculum is grounded in an indigenous perspective and makes connections to mainstream academics through indigenous approaches to learning.

Arctic

Working with Traditional Knowledge in Land Use Planning: Gwich'in Place Names, Land Uses, and Heritage Sites in the Northern Territories of Canada

Project Contributor: Ingrid Kritsch

Figure 4.15 *Hills at Tl'oondih where summer and winter trails led to traditional Gwich'in hunting grounds in the Yukon, and a clearing where one Gwich'in elder had his camp*

Credit: Ingrid Kritsch, GSCI

The Gwich'in are one of the most northerly aboriginal peoples on the North American continent, living at the northwestern limits of the boreal forest. Many families still maintain summer and winter camps outside their communities. Hunting, fishing and trapping remain important both culturally and economically, with caribou, moose and whitefish being staples of the local diet. The Gwich'in Social and Cultural Institute (GSCI) was established in 1992 because the Gwich'in were concerned about the loss of their culture and language and the impact this was having on their families. The *Dinjii Zhu' Ginjik* (Gwich'in language) is one of the most endangered Aboriginal languages in Canada. Due to the encroachment of English into all aspects of daily life, only a small number of elders and a few determined individuals continue to use the language on a regular basis, and it is rare to hear children speak the language. Government statistics in 1998 revealed that only 2 per cent of all the Gwich'in spoke the language in their home, and only 13 per cent reported they could speak the language at all. The last generation of elders who lived on the land and consequently have an in-depth knowledge of it, is passing away very quickly and there is great pressure to record their knowledge before it is too late.

The 'Gwich'in Place Names and Traditional Land Use' project (14) is carried out by the GSCI, the cultural and heritage arm of the Gwich'in Tribal Council, in collaboration with Gwich'in communities in the land claim area. The project is based in the Northwest Territories, Canada, and promotes sustainable land use among the Gwich'in First Nation through the application of their traditional knowledge to land use planning. Project research documented Gwich'in traditional knowledge and land use through the study of place names, traditional land use, ethnobotany, ethnoarchaeology, elders' biographies, genealogy, a Gwich'in language dictionary, the replication of 19th-century caribou skin clothing and the identification of National Historic Sites in the Gwich'in Settlement Region (GSR). Gwich'in place names and the associated stories along with trails, traditional camp sites, graves, historic sites, harvesting locales and sacred or legendary places are windows into Gwich'in culture and history. The project has also successfully brought elders and youth together on the land to promote and pass on the language and knowledge about the land and the culture. Efforts to record and revitalize the language are also a vital part of the work at the GSCI, where all research projects have a language component, even though it is also a challenge to find skilled people who can translate and transcribe the language in a standardized way. Funding is also a challenge. It is difficult to find multi-year funding, and consequently most of the funding for the project has been on a year-to-year basis.

Language and education programmes include language revitalization initiatives, the development of curriculum materials, a language immersion camp and an annual Gwich'in Science Camp, which is an on-the-land traditional knowledge and Western science camp for senior high school students. The inventory of heritage sites assembled during the course of this project plays a critical role in land use planning in the GSR. It is used to review land use permit applications with the goal of ensuring sustainable development of Gwich'in lands. A Gwich'in traditional knowledge policy, called 'Working with Gwich'in Traditional Knowledge in the Gwich'in Settlement Region', was approved by the Gwich'in Tribal Council in June 2004 and is used to direct all traditional knowledge research carried out in the region.

This project has proved to be very successful. The opportunity to bring elders and youth together on the land, promoting and passing on the language and knowledge about the land and the culture, has occurred during the research, which is much like the way traditional learning happened in the past. There has also been official recognition of traditional Gwich'in place names on maps and highway signage in the GSR, and the history behind these places is being acknowledged in interpretive centres in the north. One of the consequences of the work was that in 1994 one of the Gwich'in communities officially changed its name to honour the location's traditional name: Arctic Red River became officially known as *Tsiigehtchic*, a Gwich'in name that means 'mouth of Iron River'.

The project has provided information towards the production of educational materials, such as a major land-based and community history book that is now being used in local schools, an ethnobotany book and kit for the local schools, and a website that features a 'talking place name map' and virtual tours of the Mackenzie, Peel and Tsiigehtchic Rivers. It has led to the successful designation of the largest National Historic Site (NHS) in Canada (Nagwichoonjik NHS, on the Mackenzie River from Thunder River to Point Separation) and the nomination of eight Territorial Historic Sites. It has helped the Gwich'in to assess proposed land use activities in the GSR and how they might impact heritage sites. Finally, it has allowed the identification of heritage sites, special management areas and protected areas within the Gwich'in Land Use Plan, ensuring that Gwich'in lands will be taken care of in a sustainable way both today and in the future.

Aboriginal Traditional Knowledge and Assessment of Species at Risk: A Case Study from Northern Canada

Project Contributor: Nathan Cardinal

In Canada, both the inherent value and the lawful recognition of aboriginal people's traditional knowledge (ATK) are written into the Species at Risk Act (SARA). The Committee on the Status of Endangered Wildlife in Canada (COSEWIC) is the organization responsible for evaluating the status of species in Canada and is now required by legislation to base their species assessments on the best available knowledge, including both science and traditional or local knowledge. Such information has rarely been used in species conservation and the assessment of wildlife. A 2002 study of 190 reports that summarize the status of a given species at risk revealed that only one report referenced aboriginal use, and none incorporated ATK (Ellis, 2001). COSEWIC works closely with aboriginal peoples to decide how ATK will be incorporated into the process of assessing species at risk through the Aboriginal Traditional Knowledge Subcommittee. Incorporating ATK into the assessment of species at risk improves the process, and therefore the quality of designations made by COSEWIC, by bringing information and perspectives on wildlife species that are not available in published scientific literature. While extremely beneficial for species, the inclusion of ATK can more importantly signal meaningful involvement of aboriginal people in species conservation, which may ultimately improve local-level acceptance of a species' status and associated recovery programmes.

The focus of the project 'The Use of Aboriginal Traditional Knowledge in Species Assessment: A Case Study of Northern Canada Wolverines' (26) is on the importance of understanding ATK to assist the scientific community in protecting species, in this case, the threatened wolverine, *Gulo gulo*, one of the least studied of the large carnivores. This project was completed as part of a Masters' thesis, and is a research case study that investigated how ATK can be documented, described and utilized in COSEWIC's species assessment process. The study provided recommendations to COSEWIC regarding how such traditional knowledge can be gathered and utilized for future species assessments.

Wolverines are considered very important by local people, from both a cultural and a subsistence standpoint. The research found that ATK contributes invaluable information regarding the status of wolverines in northern Canada, including the special significance of the wolverine to aboriginal people, the biological characteristics of the species, relative trends in abundance, and information regarding any significant threats. ATK proved to be very beneficial for improving the validity and acceptability of species assessments. ATK from the study was found to be congruent with contemporary scientific knowledge of wolverines, supporting various studies conducted on wolverine behaviour, habitat use and food requirements. It provided finer-scaled information than currently available for many areas in the north, and further refined the present relative abundance maps for wolverines.

ATK also contributed new information regarding wolverines and clarified threats to the wolverine, especially regarding regional differences in impacts due to wolverine harvest. The study concluded that the inclusion of ATK improves the quality of species assessments to some degree and that active involvement of aboriginal people and their knowledge in the assessment process will increase the acceptability of decisions resulting from assessments at a local level. It was also noted that because of the unique cultural and historical characteristics of ATK, extreme care must be taken in its gathering in order to ensure the proper respect and acknowledgement that the knowledge and its holders deserve.

There were some challenges to the project. Not all people agreed to be interviewed, due in part to a lack of support from community organizations where people's time was already stretched, and in part to people being wary of the study. There was somewhat greater resistance to being interviewed in the larger centre, as opposed to the smaller communities. In larger communities, generally people will have less familiarity with one another, and typically are not as close-knit as in smaller communities. In smaller communities, it was easier to facilitate contacts due to the familiarity among people and their community organizations. In addition, many of the areas visited in the north were already covered by comprehensive land claims, while in the south land claims are ongoing, which usually engenders a more politically sensitive climate. This may make ATK studies in the south more difficult in some respects than in the north, due to the often controversial nature of ATK and the unsettled nature of many land claims and treaty negotiations.

The outcome of this research supports the development of long-term relationships between aboriginal peoples in Canada and the species at risk scientific community. It is expected to markedly change governmental wildlife policies.

North America

Combining Environmental Stewardship and Economic Renewal in Northern Canada: The Whitefeather Forest Initiative

Project Contributors: Alex Peters and Andrew Chapeskie

Figure 4.16 *Preparing fish in Pikangikum: people are moulded by the land and everything they draw from it, say the elders*

Credit: Whitefeather Forest Initiative

The Whitefeather Forest planning area, located in the boreal region of Ontario and Manitoba, Canada, is a holistic network of both natural and cultural features that results from the relationship between Pikangikum (Ojibwa) people and their ancestral lands. This relationship expresses a closeness that comes not only from their knowledge of using the land, but also from a spiritual and emotional connection to the land. Elders' teachings stress the importance for the Pikangikum First Nation to continue to follow the customs of cherishing the land and all living creatures, and to carry on with the responsibility of 'keeping the land'. As Pikangikum Elder Whitehead Moose puts it: 'Everything that you see in me, it is the land that has moulded me. The fish have moulded me. The animals and everything that I have eaten from the land has moulded me, it has shaped me. I believe every Aboriginal person has been moulded this way.' For the Pikangikum, the land and people are inseparable. Their territory is not merely a landscape modified by human activities, but a way of relating to the land, and a way of being on the land (Pikangikum First Nation and Ontario Ministry of Natural Resources, 2006).

The 'Whitefeather Forest Initiative' (29) combines environmental stewardship with economic renewal strategies to enable the Pikangikum First Nation to develop new resource uses, with the aim of providing urgently needed tribal enterprise opportunities for the youth within their traditional territories. The ecological richness of these territories forms a cultural landscape that is of international ecological significance – from vast tracts of jack pine to the wild rice (*manomin*) stands planted long ago by the Pikangikum people to increase food for fur-bearing and aquatic animals, to the numerous pristine waterways that flow through the forest. The cultural heritage also includes features such as pictographs, traditional campgrounds, portages and waterway channels.

The Whitefeather Forest Initiative applies a community-based land use planning approach, in which the elders of the community take a leading role in planning through a steering group. The knowledge tradition, language and stewardship values of the community guide the development of the initiative. The elders, whose knowledge and wisdom are highly valued, work with the community research team members to develop new forest-based livelihood opportunities for the youth of Pikangikum. The goal is to ensure that the maintenance of forest cover and biodiversity and the care of vulnerable species are achieved within a new economic and resource management context that also maintains the vitality and strength of the indigenous language, culture and knowledge tradition of the community. An Indigenous Knowledge Teaching and Training Centre is also a part of the project.

This approach is strengthened through innovative partnerships for education, resource management and business development with parties that have an interest in the Whitefeather Forest Planning Area. Through these partnerships, Pikangikum seek to continue their role as keepers of the land, while at the same time recognizing other interests and harmonizing them with their own interests. The Whitefeather Forest Initiative is carried out in a spirit of cooperation and mutual respect. The process to develop the Whitefeather Forest Initiative in general and partnerships in particular is centred on consensus building and dialogue-based decision making. The essence of the Pikangikum view of partnership was expressed by Elder George M. Suggashie to representatives from environmental organizations: 'We are happy when people come to us and ask how we can work together. We are very upset when things are done to our land without our participation.' The research results from collaborations with the Whitefeather Forest Research Cooperative are made available in both Ojibwa and English.

In addition, the project seeks to establish a linked network of protected areas. A Protected Areas Accord was signed in 2002, with the goal of achieving UNESCO World Heritage status. Local knowledge also played a key role in the development of a Community-based Land Use Strategy for the Whitefeather Forest and Adjacent Areas in the context of the Province of Ontario's Northern Boreal Initiative, which seeks to develop forest management approaches that are ecologically suited to the northern boreal forest.

Traditional Knowledge for Sustainability: Land use Planning among the Gitxaała of British Columbia, Canada

Project Contributor: Charles Menzies

Figure 4.17 *Traditional fishing site in Gitxaała territory showing beach at low tide, with ancient stone fish traps (semi-circular) visible in the intertidal zone*

Credit: Tristan Menzies

For many generations, the Gitxaała people have lived in their territories along the north coast of what is now British Columbia, Canada. Gitxaała laws (*Ayaawk*) and history (*Adaawk*) describe in precise detail the relationships of trust, honour and respect that are appropriate for the well-being and continuance of the people, and also define the rights of ownership over land, sea and resources within the territory. However, with the arrival of the first *K'mksiwah* (Europeans) in Gitxaała territory in the late 1780s, new forms of resource extraction appeared that ignored, demeaned and displaced the importance of the *Ayaawk* and *Adaawk* in managing the Gitxaała territories. The new industries (such as forestry, fishing, mining) have relied almost completely upon European science for management and regulation. During the last two decades, there has been a turnaround, and the value of traditional ecological knowledge (TEK), such as that reflected in the Gitxaała *Adaawk* and *Ayaawk*, has been increasingly recognized.

The project 'Forests and Oceans for the Future' (13) is a collaboration between community members from the Gitxaała Nation, a Tsimshian First Nation in British Columbia, and anthropological researchers from the University of British Columbia (UBC). The principal focus of the project is the use of Gitxaała traditional ecological knowledge for provincial government land use planning. From conception through implementation, revision and reporting back to the whole community, collaboration and respectful research practices are the central and fundamental principle of the work. Implementing this approach is not a straightforward application of rules of conduct, but instead is built upon a conception of research as a long-term relationship – which requires goodwill, commitment and compromise.

A key component of this project is to document and facilitate the deployment of customary forms of governance among the Gitxaała that regulate human action within the environment, acting to conserve and enhance biodiversity, and leading to long-term sustainability within the Gitxaała traditional territory. Policy development and evaluation is another key component of the project and involves research designed for use within the provincial government's Land Resource Management Planning (LRMP) process. Project team members contributed to preparing and presenting reports on the Gitxaała informal economy and TEK for use in the North Coast LRMP. Public education materials have been developed to facilitate the sharing of knowledge and understanding of the issues, controversies and concerns related to forestry and natural resources. These materials were inspired by the experiences of students and community members living within the Tsimshian territory of British Columbia. Seven papers based on Forests for the Future Research were published in Volume 28 (1 & 2) of the *Canadian Journal of Native Education* and are available online (www.ecoknow.ca/journal/index.html). These papers are one of the outcomes of this unique collaboration between anthropology researchers from UBC and community members from Gitxaała. Project results suggest that rather than to macro-level planning authorities, resource management should be devolved to local-level organizations in conjunction with non-aboriginal people living within the territories. Individuals, agencies or corporations based outside the region should have a restricted role and limited access to resources and decision-making authority in the local arena.

Recovering the Connection between People and the Environment through Ancestral Law in British Columbia, Canada

Project Contributor: Patricia Vickers

The Nisga'a People of the Nass River have lived on the northwest coast of British Columbia, Canada, for generations – long enough for a culture to thrive, adapt and endure. For the Nisga'a Nation, the meaning of the relationship between people and the environment is found in metaphor and stories. This long-held connection has been undermined by the long-term effects of colonization (in which residential schools played an enormous role) and unsustainable development choices such as fish farming and clear-cut logging, because of the immediate need to alleviate poverty. This has tended to undermine the development of initiatives that honour and revitalize culture, such as cultural centres and retreats, and programmes for children, community members and tourists.

The project 'Transforming the Cage' (38), supported by the Laxgalts'ap Village Government, aimed to identify the roots of an internalized sense of inferiority that affects the Nisga'a, due to the history of oppression from colonization, and the impact that this has on daily living. *Ayuuk* (ancestral law) is promoted to deal with conflicts in family and business relationships. The *Ayuuk* holds the knowledge of rites of passage, protocol for marriage, birth and death, and resolving conflict, and guides the Nisga'a in creating spiritual balance in a reciprocal relationship with the environment at both the individual and the collective levels. The Nisga'a Lisims Government – a modern administration that draws from traditional culture and values – has worked with the Nisga'a Nation to build a culture and economy that respect and protect the Nisga'a natural and cultural heritage. In the words of the Nisga'a Lisims Government (www.nisgaalisims.ca/?q=welcome), today the Nisga'a Nation is a place where 'our Ayuuk, language, and culture are the foundation of our identity; learning is a way of life; [and] we strive for sustainable prosperity and self-reliance'.

Supporting Traditional Health Practices in Urban Areas: Indigenous Theory for First Nations Health in Canada

Project Contributor: Dawn Marsden

The dissertation project 'Indigenous Theory for Health: Enhancing Traditional-Based Indigenous Health Services in Vancouver' (2), completed in 2005, was supported by the UBC and by grants from the Canadian Institute of Health Research (CIHR)-funded BC Aboriginal Capacity and Developmental Research Environment (BC ACADRE). It was developed from the informal recommendations of traditional indigenous practitioners. It aimed to address the health impacts of colonization and subsequent discontinuity between migrating indigenous peoples and their traditional territories, by raising the idea of supporting traditional health practices in urban areas. A team of 22 traditional-based practitioners, facilitators and clients explored the challenges, opportunities and recommendations for revitalizing traditional health teachings and practices among urban indigenous populations in Vancouver, British Columbia. This study was unique in its exploration and application of indigenous theories, methodologies and methods (holism, indigenous protocols, dreaming, prayer and talking circles) to health service research. As a member of the traditional research group stated in 2004: 'We not only need to have our own health care, our own dental clinics, we need to have a place where our people can possibly be treated respectfully. But we don't have that. We don't have our medicines, we don't have our Elders, and ... we need to have a gentle place to heal.'

The results of this study can be summarized as a collective determination to establish an inter-Nation council of practitioners, under the umbrella of local land-based Nations, for the development of ethical guidelines and standards for practice, apprenticeship, communal resources, professional development, referral and community outreach; and to raise the status of traditional practices, while reducing racism and negotiating for traditional health services with provincial and federal governments. These recommendations called for protection of local traditional medicine harvesting sites and sacred practice sites, and the development of appropriate environmental space for holistic healing, with essential inclusion of clean water, fire, earth and air.

An underlying principle of this project was that revitalizing lifestyles based on a deep reverence for the interconnectedness between humans and the environment will foster balanced living, thus influencing a societal shift toward more sustainable practices. One focus of this study was the transmission of indigenous worldviews, which are seen to arise from multi-millennial sustainable relationships between specific humans, plants, animals, waters and lands. These worldviews contain whole knowledge systems, embedded in language, values, practices and material goods, which – when intact – produce ecological and socio-cultural resilience to adversity and conservation of biological diversity. The transmission of such traditional knowledge systems is seen as vital to the maintenance of sustainable cultural continuity and bioregional management systems. These systems are renewed throughout the life cycle, through health-related spiritual teachings and ceremonies (e.g. birthing, coming of age, dying) that reinforce indigenous identification with Mother Earth and all the beings living upon her.

The main challenge following this study has been that its recommendations are at risk of being forgotten. While any efforts to implement the recommendations can be facilitated by others, the development of an inter-Nation council must be led by traditional indigenous practitioners. In support of this process, the project contributor has applied some of the concepts to current First Nations' health issues through a CIHR-funded postdoctoral fellowship, contract research with an aboriginal women's research group, and research with a national-level First Nations health organization in Canada. Various community pilot studies and knowledge translation exercises have been conducted in the areas of visioning, food security, injury prevention, violence prevention, practitioner recruitment, retention and remuneration, strengthening families, gender-based analysis, mentoring and research methodologies. At the same time, the integration of traditional indigenous principles and practices has been intensified through activities across Canada to develop cultural competency and cultural inclusion in health services, through consultations with, and employment of, indigenous peoples.

Learning that Wisdom Sits in Places: Apache Students Reconnecting to Land and Identity in Arizona, US

Project Contributor: Jonathan Long

Figure 4.18 *Apache students identifying plants at Goshtlish Tú Bil Sikąné*

Credit: Jonathan Long

Over 25 years ago, nearly 300 places of cultural importance to the Apache people in the valley of Cibecue, Arizona were mapped and photographed by anthropologist Keith Basso with the help of Apache tribal elders. The results were published by Basso in 1996, in a book called *Wisdom Sits in Places* (Basso, 1996). Many years later, in 2005, students at the Cibecue Community School initiated the project '*Ndee bini' bida'ilzaahi*: Pictures of Apache Land' (39) with several objectives. First, the project was to teach the youth in the community about traditional Apache values for the land. This was accomplished by identifying the Apache names of places and finding the stories that go along with them and tell the students the historical, social and moral interpretations their ancestors had of these places. Second, the aim was to combine traditional ecological knowledge with the scientific method to explain the changes in the land. Third, the students had to analyse the changes in the environment from a personal and social perspective. Finally, the aim was to instil in the youth a commitment to restoration of their land and waters. Funding originally came through a number of different sources, mostly US federal grants. The summer project in Cibecue was started using an environmental education grant from the US Environmental Protection Agency as well as support from the Cibecue Community School.

White Mountain Apache culture emphasizes the infusion of the physical world with mental and spiritual dimensions. The Apache language illustrates the inseparability of the two: for example, the root word *ni'* can either refer to the 'mind' or to 'land'. Places within the landscape remind people how to live right (Basso, 1996), and people's behaviours affect the conditions of the landscape. Water bodies hold exceptional significance, as nearly half of the place names in many regions of aboriginal Apache lands are associated with water bodies or wetland species (Grenville Goodwin Placenames Project, 1997, cited in Long et al 2003).

The largest fire in the history of the Southwest, the Rodeo-Chediski wildfire, which occurred from 18 June to 7 July 2002, struck Cibecue with a tremendous impact. The wildfire provided the impetus for the project to restore the springs and wetlands that were damaged. At the same time, there was a need to better engage the divided community of Cibecue in restoration research and planning. The students visited 16 of the original sites that Basso had been to, took photographs, and conducted an inventory of the plants, rocks, soils and water. They also conducted interviews with their elders to better understand how the land has changed over this time period. They compiled their findings in a computer database, including the Apache names for plants, places and other ecological features, and prepared a poster, slideshow and video to share their findings with community members. As the project is ongoing, the students will also prepare an exhibit for the tribal museum based on these findings. The students have worked on two ecological restoration projects, and future plans include working with community members to plan more restoration projects for additional sites that they have studied. Extensions of the project may include recording information needed to safeguard springs and aquifers from drawdown by groundwater pumping, and to guide the protection and restoration of areas damaged by wildfires. The project has led to significant investments in post-wildfire monitoring of springs and several rehabilitation/stabilization projects. These projects are an important step forward in expanding the scope of the federal post-wildfire response effort to better address impacts on eco-cultural resources. The programme has also been talked about as a possible model for other communities on the Reservation.

An important part of the work is reviving pride and identity among the youth of Cibecue. In the process of gathering data (soil, plants, water, geology, GPS, etc.), the students learn why their ancestors held such respect for water and reverence for these sacred places. This learning is especially important now when young people are losing their language and identity, and assimilation is taking place because of modern-day technology and lifestyle. By learning the Apache names for features of the land in their own backyard, they understand that place names speak to the land and its attributes, as well as the condition of the land and the traditional values of their people. There is immeasurable pride that comes with true understanding of one's culture, a feeling of reverence and the appreciation of why the land and water are sacred and how the land still speaks to the morals and values of the Apache people. This, the project leaders feel, is 'what makes our project unique'.

So far, the students are learning from the land – as they listen, observe and study, they hear the springs speak to them and they understand that the water is sacred. At the same time, they understand the land from a scientific perspective as well. The community is also beginning to understand the nature of the project, and people are starting to provide input by giving additional information on what they know of the changes that have taken place in their lifetime and making recommendations as to what they think is important for the youth of the community to study. For example, they want the students to learn about medicinal plants. One elder said she is willing to teach someone how to boil medicine for healing. Another elder said he knows of where there is a hot spring and he would be willing to show the students. Yet another elder gave a story of a lake and a spring, which she said is located on an old trail that was used by the old ones. The students have demonstrated deeper cultural knowledge and a greater willingness to speak in Apache and develop proficiency in the language. 'The names of all these places are good. They make you remember how to live right, so you want to replace yourself again' (Nick Thompson, quoted in Basso, 1996, p59).

Latin America

Recovering Landscape Health and Cultural Resilience in the Sierra Tarahumara, Mexico

Project Contributor: David J. Rapport

Figure 4.19 *Luís and Tomás Palma gathering native pine seeds for a tree nursery in their Sierra Tarahumara community*

Credit: David J. Rapport

The Rarámuri people (also known as Tarahumara by non-Rarámuri) are an indigenous group living in the Sierra Tarahumara, a part of the Sierra Madre Occidental mountain range in the northern Mexican state of Chihuahua. This region of high sierras and deep canyons boasts an exceptional ecological diversity, and is home to some of the most resilient indigenous societies in the North American continent. The Rarámuri (about 70,000 people, living mostly in isolated settlements and small villages scattered across the Sierra Tarahumara) speak a distinct language and have maintained a strong identity and vibrant cultural traditions through over five centuries of contact with the now prevailing Spanish-speaking population. They are subsistence farmers, and have traditionally also relied extensively on a variety of wild plant and animal species for food, medicine and other basic needs.

However, their long-term adaptation to this mountainous region and their ability to sustain their livelihoods and way of life – and ultimately to retain their cultural and linguistic identity – have been severely threatened by rapid environmental, socio-cultural and economic changes brought about by virtually unrestricted mining, logging, ranching, mass tourism and now increasingly the drug trade, all of which have been facilitated by extensive road development and the building of other major infrastructure. These activities have collectively resulted in massive deforestation causing the loss of forest plant and animal species; over-grazing; soil erosion with consequent loss of water resources; frequent droughts and flash floods; water pollution; decrease of arable lands and diminished soil quality and fertility, resulting in lower crop yields and periodic crop failures; displacement from traditional lands; out-migration, especially of the younger generations, due to inability to make a living in the communities; induced social and cultural change; social dislocation and loss of social cohesion; erosion of intergenerational transmission of values, beliefs, knowledge, practices and language; and a variety of health and nutritional issues. Adding to these woes, global warming is projected to bring long-term drought to the region.

The scale and pace of change are challenging the Rarámuri's ability to continue to live and develop according to their own worldview and way of life. Many elders and other community members are concerned about the Rarámuri's future as a distinct people if the erosion of their landscape and culture continues. While stressing their long-standing resilience as an indigenous people, they perceive threats to their physical, cultural and spiritual survival and to the transmission of Rarámuri identity, values, knowledge, customs and language to younger generations. They see the need to take action, and some of them recognize that, in addition to their own efforts, they can potentially benefit from working with outsiders who can provide needed expertise and other resources.

In response to this need, the project 'Eco-cultural Health in the Sierra Tarahumara, Mexico' (6), spearheaded by the NGO Terralingua with funding from The Christensen Fund and Canada's International Development Research Centre, was developed in partnership with two Rarámuri communities in the vicinity of the town of Norogachi. The project began in 2006, building on a relationship between Terralingua and the Rarámuri that had been evolving since 2000. At a meeting between Terralingua and traditional Rarámuri authorities, elders and youth in 2004, consensus had emerged to work together on a collaborative project focused on the recovery of the health of the landscape and the social and cultural resilience of the communities. Participants agreed that the first priority should be water, which they saw as the basis for all life and at the same time as an increasingly scarce and unhealthy resource, with serious consequences for humans, animals, forests, wild plants and crops. Revegetation was also a high priority, along with concerns about human health and culturally appropriate education for children and youth. After the authorities consulted back with their respective communities, an official invitation was issued to Terralingua to come back to work with Rarámuri communities.

According to the priorities expressed by the Rarámuri, the project was conceived in phases. While the ultimate goal was the development of an on-the-ground, practical education programme that would assist the Rarámuri in their effort to recover and take direct control over the eco-cultural health of their landscape and communities, initial steps focused on bringing in potable water to one of the two participating communities, developing tree nurseries and home gardens, assessing issues of health, hygiene and sanitation, and addressing literacy for women. For these purposes, Terralingua formed an interdisciplinary team of collaborators with expertise in biocultural diversity, ecosystem health, human health, hydrology, ecological restoration and indigenous education. Project activities in this phase took place between 2006 and 2008, with five field visits by Terralingua team members, while community members continued activities between visits.

The potable water project was carried out entirely by community volunteers. The Rarámuri had already identified a distant upland spring (about 8km north of one of the two settlements) that has good drinking water. They had the intention of bringing water to one of the two communities, which had no potable water, where the people had had to resort to drinking polluted water from a nearby stream and various seeps and pools. For this purpose, they had previously built a small holding tank there, but lacked the resources to lay a pipeline from the spring to the community. With materials provided by Terralingua, the volunteers undertook the project, which involved not only an important engineering aspect (laying the pipeline and burying it in places over rocky ground), but also building an unusual level of cooperation among several settlements along the route. This effort required taking time off from daily subsistence activities and was completed over a period of about one year, as allowed by weather conditions (summer floods, winter ice) as well as seasonal farming needs (planting and harvesting) and the occasional need to earn income in the off-season by working outside the settlements. At present, the entire pipeline has been laid and the system is in operation. Community members have also taken the initiative to build a large holding tank near the settlement for long-term water storage, as a way of countering the effects of the dry seasons and the periodic droughts.

An assessment of the health of the local landscape showed considerable evidence of degradation – owing to the combination of massive deforestation by outside logging interests and over-grazing by both non-Rarámuri cattle ranchers and Rarámuri farmers. Much of the landscape around the settlements has lost most of its topsoil, with large erosion gullies visible everywhere due to the action of winds and rains. The complexities of landscape-level restoration were compounded by lack of secure land tenure and control over land use around the settlements. It was apparent that community members could not always effectively control land beyond the immediate vicinity of their household compounds, due to incursions by unfriendly neighbours and cattle ranchers.

Initially, the project entertained the possibility of starting pilot revegetation on a hillside identified by community members, which would have been fenced off to keep out the grazers (both the goats owned by the Rarámuri themselves, as well as the larger cattle, often owned by non-Rarámuri who take them to pasture in the area). Some of the restoration techniques that would have been applied, and that were demonstrated at the outset (such as creating swales by laying rocks and branches across the slopes to impede water runoff and capture soils) were actually akin to the traditional Rarámuri practice of building *trincheras* (ditches) along hillsides – a practice that some of the local elders mentioned, but knowledge of which seemed to have disappeared or have gone dormant among younger generations.

However, doubts soon arose that it would not be possible to adequately protect this site from grazers long enough for revegetation to take hold. Therefore, the consensus was that it would be better to start by establishing tree nurseries near the households, over which people could have greater control. Community members would then be able to transplant trees close by, where there is little or no shade or plant life, with the added possibility of selling seedlings in the nearby market town. Transplanting nearby would also provide an easily accessible source of firewood, whereas people currently have to go long distances to the remaining wooded areas to provision themselves with dead wood and fallen branches.

A small temporary nursery was set up, and people gathered and planted the seeds of local pines and oaks in improvised containers made from plastic bottles and tin cans filled with topsoil from the nearby riverbed. Some nut trees, such as walnuts, were also planted as a source of commercially viable fruits. In addition, this provided an opportunity to demonstrate the preparation of a compost pile using plant materials and manure for later use to fertilize the fields – another practice that, according to elders, was germane to traditional practices, and had probably been supplanted by the introduction of chemical fertilizers. Subsequently, with materials and guidance provided by Terralingua, community members built a fully fledged enclosed tree nursery, which was enriched with soil from the riverbed and in which four kinds of local pines were planted. The nursery includes an irrigation system with half-inch pipe and a hand-held sprinkler engineered by the community, with which they can readily water the plants. The results so far have been mixed, in part due to extended drought, which has threatened the viability of the seedlings, and in part due to some of the seeds gathered locally (particularly oaks) failing to germinate. Most of the pines and walnuts have been growing, however. Eventually, the nursery might supply pine seedlings for hillside revegetation wherever possible. Aware of the role of trees and other plants in holding soil and moisture, community members also intend to do some tree planting near the upland spring, to help preserve their water, and to plant agaves around other smaller water sources to retain both soil and water.

Terralingua team members also carried out a survey of health, hygiene and sanitation issues, practices and concerns in the settlements, with the goal of incorporating these topics in the later development of an educational programme from an eco-cultural health perspective. The survey was coupled with demonstrations of hand washing and sanitary handling of food and water. Further, the survey sought to assess nutritional status, as evidence from the medical literature suggests that a shift from the traditional Rarámuri diet, consisting largely of corn and beans, toward an increasing adoption of non-indigenous foods is responsible for negative changes in Rarámuri health. This has combined with malnutrition due to periods of drought and other effects of environmental change on soil fertility and crop abundance. In order to help improve food supply and nutrition, the project worked with community members on home gardens, demonstrating various techniques for capturing rainwater and grey water for irrigation, creating contours to retain water and soil, using mulch, and increasing the production of vegetables and fruits. Two enclosed home gardens were built, with cooperation between families that did not normally work together. The intent was that each family would harvest the food, but share the seeds with the community, thus enhancing community interaction and cooperation and reinforcing the kind of community solidarity that is indispensable to strengthen cultural identity and support cultural affirmation.

Along with the health survey, project team members surveyed the situation of educational services in the two communities, to assess existing education programmes for children and adults. Existing programmes mostly follow a conventional transitional bilingual education approach, that is, one that only uses initial literacy in the indigenous language as a stepping stone for literacy in Spanish, after which literacy in Rarámuri is no longer maintained. Community members themselves appear to generally favour literacy in Spanish as a means to better navigate the outside world, and tend to attribute lesser value to literacy in Rarámuri. In the course of the survey, in fact, community women expressed the desire to learn to read and write in Spanish. Some literacy sessions were conducted, during which the women learned to recognize and write their names. Because the prevailing educational approach disfavours or altogether excludes Rarámuri language and culture, and often forces children to travel a long distance (mostly on foot) to go to day school, or even to attend residential schools, the project aimed to determine whether and how Rarámuri language, culture and traditional knowledge might be integrated in alternative in-situ education initiatives within an eco-cultural health framework. This goal dovetailed with the interests of some of the Rarámuri (including an influential elder), who are more keenly aware of the threats that the existing educational system poses for the maintenance and intergenerational transmission of Rarámuri identity, language and worldview, and for community cohesiveness.

Based on these experiences, the second phase of the project aims to focus on two goals: the development of hands-on eco-cultural health educational materials with and for the Rarámuri, intended for elementary school children and youth while also crucially including teachers and community adults and elders; and capacity building for adult community members to carry out further ecological restoration and improve landscape and community health. The guiding philosophy is the co-creation of knowledge, know-how and educational materials, bringing together traditional and scientific knowledge and with a community-oriented, service-learning approach. It is also clear that, for a long-lasting impact, it is imperative to go beyond what can be accomplished during the project team's time-limited visits to the communities, by creating the means for continuity and self-sustainability of educational and on-the-ground activities. A focus on 'training the trainers' (selected community members, teachers, health providers and others involved in community-level work) is central to accomplishing this goal in the longer term. At present, the main challenges for undertaking this phase of the project reside in mustering new funding during a global economic downturn, as well as an increase of tension in the region due to escalating drug-related violence. A hopeful sign, however, is that a local and an international foundation are working together to establish an overall Rarámuri education initiative in the Sierra Tarahumara, which shares many of the project's goals. Terralingua may be able to join forces with this larger initiative and link it to the communities with which the project has been working, thus ensuring that they benefit from the initiative.

One of the key lessons learned in this project has been that entraining and sustaining a truly participatory community process is a long-term and complex undertaking. This is particularly the case in a situation in which the ecological, socio-cultural and economic issues involved are on a scale far larger than those local people may traditionally have had to contend with, and that require a level of community cooperation far greater than usual – while at the same time those very issues pose immediate survival challenges that community members are often led to confront individually rather than cooperatively. Some of the Rarámuri fully realize that the scope of the threats they are facing requires them to go beyond individualism (and sometimes community rivalries), and they strongly advocate working together to address the problems. From this point of view, it appears that the project has had a positive role, facilitating a number of community discussions and reflections on the issues at hand that would rarely have happened otherwise, and fostering collaborative work that people might not have engaged in otherwise. As one leading elder put it, in expressing his satisfaction for this process and exhorting his community to continue along this path: 'It has been an awakening for us.' The 'awakening' is still tenuous, however, constantly challenged by the forces that are bringing about rapid ecological, economic and socio-cultural change. If the project succeeds in further developing its educational and capacity-building activities, more enduring seeds for eco-cultural survival may be sown.

Strengthening Indigenous Cultural Heritage through Capacity Building in Costa Rica

Project Contributors: Hugh Govan with Rigoberto Carrera

There are eight indigenous groups in Costa Rica, numbering some 63,800 people, which comprise 1.7 per cent of the national population. Half of them are now settled in 24 reservations or territories, which cover an area of approximately 325,470ha or 6.3 per cent of Costa Rica. The indigenous groups are: the Cabécar, Bribri, Brunca or Boruca, Térraba, Huetar, Guatuso or Maleku, Chorotega and Ngäbe-Buglé. In 2001, two new reservations were created by law: Altos de San Antonio (for the Ngäbe-Buglé) and China Kichá (for the Cabécar).

The Ngäbe people number some 180,000, principally located in Panama, although around 4000 reside in southern Costa Rica, close to the Panama border. The Ngäbe-Buglé of Costa Rica inhabit five reservations or territories in the south of the country: Coto Brus, Abrojos Montezuma, Conte Burica, Altos de San Antonio and Guaymí de Osa. The 23,600ha of Ngäbe reservations maintain around 70 per cent forest cover, consisting of a rich variety of habitats encompassing three of the five elevational zones found in Costa Rica (tropical, premontane and lower montane) and three of the four humidity provinces (rain, wet and moist). Examples are the tropical very wet forests of the Osa Reservation and the lower montane moist forests of Coto Brus.

The Costa Rican Ngäbe are among the poorest people in the country, but until recently there were almost no development initiatives taking place in their territories. This is due in part to difficulties in funding and cash flow problems. In part, it is also due to their legal status: the Ngäbe were not accepted as equal-rights Costa Rican citizens by Congress until 1993. The Ngäbe face a variety of major challenges, including the occupation of up to 25 per cent of the reservation's area by non-indigenous settlers, poor access to health services and limited options for the production of food and cash. The abysmal indicators for all these problems are at odds with the generally good quality of life experienced by the majority of Costa Ricans.

Using a co-management approach in collaboration between the Ngäbe people and the NGO Fundación TUVA, the project 'Support Project for the Ngäbe Indigenous People (Proyecto de Apoyo al Pueblo Indígena Ngäbe)' (20) was set up to strengthen the organizational capacity and leadership of the Ngäbe, in order to reverse the loss of their culture, recover traditional political institutions and traditional medicine, support territorial defence and appropriate management practices, and improve agricultural production systems, health care and education. Initially, the project's emphasis was on sustainable production and use of traditional medicines, based on the priorities of the Ngäbe people, who determined what a healthy and vibrant community meant to them.

The project has supported a council of traditional healers who have operated a successful apprentice programme and worked with a team to produce a book of traditional plants used for healing. This book was produced in the Ngäbere language and only distributed to the healers, owing to concerns regarding issues of intellectual property rights and benefit sharing. Responding to calls by the Ngäbe for the recovery of oral history and the teaching of it in indigenous schools, a team of youths and elders became involved in the production of the first volume. Inspired by this initial effort, the Ngäbe youth have continued with the production of a second volume and also a more ambitious project, a CD and tape of traditional songs. This involved coordination between elders and the youth who taped and transcribed stories, as well as with the Ministry of Public Education and teaching staff. These efforts were recognized by the nationally prestigious Ford Motor Co. Conservation and Environment Prize for Cultural Heritage awarded by the Minister for Environment. In the process, the Ngäbe youth are learning to write in their original language and a book on traditional medicinal plants has been written in the Ngäbere language. Elders, indigenous teachers and Ministry of Public Education all contributed to establishing the written standard for the language.

A guidebook was also produced that interprets the legal rights of the Ngäbe to defend their territory and resources, and claim their land rights. The project included a legal study to influence policy change with regard to indigenous rights to manage natural resources. The project was carried out in collaboration with the Amazon Conservation Team, which provided funding along with the UNDP Small Grants Programme, Fundación CRUSA, IUCN, Fundecooperación and the Embassy of The Netherlands. The project was completed in 2003–2004, but some activities are ongoing and some participants are now local leaders.

Reviving Traditional Seed Exchange and Cultural Knowledge in Rural Costa Rica

Project Contributor: Felipe Montoya Greenheck

In Costa Rica, agrobiodiversity has been lost because of market pressures on agricultural production. The demand for high-volume, standardized production has been a disincentive for the continued cultivation of low-yield traditional seeds, even though the traditional varieties have for generations been selected for their higher nutritional value and their adaptations to local conditions. State policies promoting agricultural 'development' have provided incentives in favour of monocropping. Findings show that after only one generation of farmers not planting their traditional seeds, many of these varieties have disappeared, along with the genetic material and the associated cultural knowledge.

More recently, a new sensitivity toward biodiversity and appreciation for diversity in itself, as well as the increased cost of chemical fertilizers and pesticides, have fostered an interest in organic farming and in recovering traditional seeds, exchanging them and sharing the related knowledge. The recovery of native and local seeds is also an important link in the process of safeguarding the family farm as a way of life. The family farmer, or *campesino*, is one of the foundations of Costa Rican national identity and worldviews. The production of the family farm is the source of Costa Rican national, regional and local cuisines, along with the accompanying vocabularies.

However, the transition process from conventional to organic farming was hampered by the lack of local, traditional seeds. The umbrella organization COPROALDE, which brings together a number of Costa Rican NGOs dedicated to alternative development projects, especially involving organic farming, was not addressing this deficiency due to other priorities. That led the project contributor in the late 1990s to establish another organization, MILPA Inc., dedicated specifically to promoting the recovery of practices that would safeguard the presence of viable traditional local seeds.

The project 'Participatory Genetic Improvement of Traditional Crops and Native Tree Species' (18), supported by MILPA Inc., helped to revitalize the traditional practice of seed exchange and the associated traditional knowledge among Costa Rican small farmers. Although the project ended several years ago, and MILPA stopped being active as an organization, the network of seed exchangers that the project promoted continues to grow, and is helping to build an organic farming movement based on diverse, locally adapted organic seeds. Valuing this local genetic diversity is helping rekindle appreciation for the local knowledge that previously had been cast aside as worthless. Youth are also actively involved, and project information is included in studies at the local university. The project illustrates how biocultural diversity conservation is linked to landscape conservation, to alternatives in sustainable development, and to the quality of life in general. In the words of the project contributor: 'Biocultural diversity is our last resource pool that we need to maintain. It is the non-fossil fuel that will keep the world rich in many ways.'

Furthermore, not only has COPROALDE taken on the ideals of the project and established a weekly organic farmers market with the exchange of seeds as a central feature, but also the Ministry of Culture and the Ministry of Health are currently collaborating in a nationwide project aimed at protecting food traditions and sub-utilized foods. With looming threats to food security throughout the world, the need to secure national food production and local and native seeds becomes an issue of national security. Protecting cultivated biodiversity is fundamental for the survival of Costa Rica and the cultural diversity within it.

Tejedores de Vida: Revitalizing Indigenous Identity and Nature-Based Knowledge in a Muisca community, Colombia

Project Contributors: Gabriel Nemogá with Carlos Mamanché

The Muisca people, living at altitudes between 1200m and 3200m above sea level in the valleys of the central region of the Andean mountains in the northeast part of South America (the savannah of Bogotá, Colombia), were so named by the Spanish conquerors. The Muisca people's existence was disrupted by the arrival of the Spanish invaders, as their territories and resources were pillaged and exploited, their sacred sites looted for gold artwork, and traditional burial grounds desecrated in order to rob the personal gold and emerald possessions of the murdered chiefs. Indigenous Muisca territory was divided up in order to isolate the indigenous people into small land areas. The Spaniards imposed a territorial system of control that allowed them to appropriate large tracts of land called *encomiendas.* The colonizers and the church confiscated the most agriculturally productive lands and exploited indigenous people as cheap and expendable labour. Men were forced to pay tribute to the Spanish Crown and to provide free labour for the *encomenderos*, while women were subjected to domestic work in the *encomiendas* and often endured sexual violence. The indigenous population was devastated by the new diseases brought from Europe, genocidal policies, over-exploitation and the disruption of their social, political and economic organizations and networks. In the ancestral territory of Sesquilé (an indigenous town established by the Spaniards near the sacred lake of Guatavita), the Church confiscated the lands of indigenous peoples from the end of 18th until the mid 19th century. The Muisca people from Sesquilé were gradually pushed into higher elevations and the more marginal mountainous regions. As recently as the mid 1970s, the municipal authorities were appropriating indigenous territories by breaking up the *resguardo* (indigenous collective lands once recognized by the Spanish).

In 1991, a constitutional reform passed by the government of Colombia, with the direct participation of the indigenous delegates in the National Constituent Assembly and the support of other political parties and coalitions, acknowledged the country's cultural and ethnic diversity and gave political and legal recognition to indigenous peoples, enshrining indigenous political and social autonomy and territorial rights. This reform notwithstanding, in practice local authorities continue to disregard indigenous rights – for example, by not including the Muisca community or consulting with them in relation to their 2007–2008 territorial planning. At the national level, the Colombian government has abstained on the United Nations Declaration on the Rights of Indigenous Peoples. However, with the 1991 constitutional guarantees, different communities emerged and openly began their cultural and ethnic recovery and affirmation.

The current social, economic and political organization of Muisca communities is the outcome of their struggle to revitalize and rebuild their culture and identity. Among them, the indigenous community group 'Los Hijos del Maíz' ('The Children of Corn') in Sesquilé developed and strengthened their social, economic and political processes under the remarkable guidance of the traditional healer and spiritual, political and cultural leader Carlos Mamanché. One of the most important territories with a record of Muisca settlement is the nearby Lake Guatavita. The legend of 'El Dorado' is sourced to this lake, which has a central place in the history of the Muisca people. It was believed that the lake held immense treasure troves of precious metals. According to legend, the Muisca *caciques* (chiefs) would, during ceremonies, offer their gold adornments to the spiritual deities who inhabited the lake. Today the *hijos del maíz* families are living on land that was previously part of the original indigenous *resguardo*. However, of the 24 indigenous families interviewed in 2006, only 46 per cent had their own homes while the rest had to rent places or live with relatives (Fundación Hemera, 2006). In 1998, the community managed to purchase a 600m^2 piece of land with their own resources. The community has reintroduced traditional agricultural crops and practices and has begun to revitalize traditional weaving and pottery, the Muisca language and cultural teachings. Under the guidance of Carlos Mamanché, and with the cooperation of people from other Muisca communities of the savannah of Bogotá, a *cusmuy*, a communal meeting and ceremonial house, was erected. Since then, the *cusmuy* has become the epicentre of the Muisca community, a place for cultural revival and collective work activities for men, women and youth, as well as a place for ceremony and spiritual cleansing using traditional medicines and plants. The construction and the structure of the *cusmuy* symbolized the centre pole for the recuperation and affirmation of Muisca spirituality, thought and identity. For the community, it is the ceremonial site to dialogue with the spirits and ancestors.

The 'Weavers of Life (*Tejedores de Vida*)' project (36) was established in 2001, with initial support from regional governmental institutions and then from some Spanish NGOs such as the Farmers' Union of Catalonia (*Unión de Agricultures de Cataluña*) and the Spanish Farmers in Solidarity (*Agricultores Solidarios de España*). This funding allowed for the development of diverse economic activities and small projects, among them egg farming, conservation and the raising of deer, weaving blankets and tablecloths, wool knitting and glass beading. However, the project had deeper spiritual and cultural objectives: to revitalize and affirm cultural identity and ancestral cosmological knowledge and spirituality that would otherwise be at risk of disappearing. The community has sought to restore traditional practices, teachings, knowledge and understandings rooted in the natural world. It has worked on the recognition and conservation of the local flora, fauna, water sources and sacred sites and on the recovery of traditional food crops, native seeds and craft activities, and has developed a culturally appropriate education curriculum. It has also sought to recover traditional medicinal plant knowledge and their uses and to promote the establishment of medicinal home gardens. The revival of the use of medicinal plants has had a critical role in the affirmation of cultural identity in the Muisca community of Sesquilé.

Legal recognition of the Muisca community in Sesquilé was officially obtained from the national government in September 2006. This important legal victory was celebrated by the whole community, as it confirmed the validity of this indigenous struggle. The subsequent years, however, have seen significant challenges arise from within and outside the community. Various disagreements arose between some community members and the leaders of the cultural affirmation movement, who sought to take on harder challenges in order to consolidate the Muisca community. Also, local authorities and private landowners became concerned with the increasing strength of Muisca identity, and stalled community activities aiming to recover their sacred sites.

A more serious challenge to the project, however, came with the untimely death of Carlos Mamanché in 2007. The loss of this leader gravely affected not only the project, but also the Muisca cultural revival movement as a whole. Some project activities were suspended for a while as the community recovered from the loss. Ultimately, community members continued on and even participated in local cultural events, thus demonstrating strong community resilience. At the same time, the death of Carlos Mamanché left a leadership gap in the community, as he was one of the main knowledge holders of Muisca thought and cosmovision. He had tried to forge a core of young people to keep the process going after him, but his premature death caused some youth to abandon the movement. Only two members had begun training in traditional medicinal knowledge, while no members had engaged in mentorship around social and political organization. The organizations supporting the community process were forced to change their emphasis, shifting from biocultural activities, to providing legal assistance and technical training to community members who have taken on leadership roles.

The current Directive Council has proposed to improve family food security through home gardens, to revive ancestral practices for sustainable agriculture. This involves community members, especially women who have received technical training from governmental entities. The main limitation for families, however, is the lack of land to cultivate. According to the community census, there are 156 families and only 10 are active in this agricultural project (interview with Rafael Mamanché, Council President, 2009). The Council decided to focus the agricultural activities on the collective land bought in 1998. The Council and the community as a whole face economic difficulties, as people do not earn enough income to devote themselves to being full-time community leaders. Some outside organizations and individuals have come to develop projects, such as ecotourism, but without respect for indigenous integrity and dignity. For example, a private-sector business venture proposed a 'theme park' development, in which the indigenous peoples' role was to be solely as tourist attractions, dressing up as 'authentic' Muisca Indians. 'We are facing economic difficulties but we are not for sale' said the council President. 'We cannot let go of what Carlos built with such sacrifice and effort; mainly his teachings about what it means to be Muisca, our identity and dignity. We had a commitment with our children and with the communities that supported our struggle for better life for indigenous peoples in Colombia and the respect of our rights.'

In spite of the setbacks, the community has been able to continue important activities with children. With the assistance of NGOs, private schools and universities, the Muisca have organized visits to the community in exchange for small monetary honoraria. This earned income goes directly and exclusively to support children's activities. Sometimes visitors provide workshops and/or handicrafts activities instead of cash. Musical training for the youth was halted when the former leader left the community, but new resources have been allocated for musical training for younger children. With economic support from the Spanish Foundation Payesos, the community is building a place for young people to become involved in new projects and initiatives aimed at youth.

The community is also working on a new activity funded by the environmental governmental organization Regional Autonomous Corporation (CAR), which promotes nature walks through traditional territory. The activities include elders' teachings, an emphasis on community life and discussions about management of collective lands. The community in Sesquilé shares its work on medicinal plants, traditional and spiritual revitalization with other surrounding Muisca communities. The community is self-sufficient in traditional medicine through management of wild medicinal plants, although the adverse circumstances of 2007 limited the potential activities of such projects.

Traditional medicinal practices are also being reclaimed, including the healing visions, the sweat lodge (*temazcal*) and the use of sacred plants like tobacco and *yagé* in ceremonies. Ceremonies are organized for the community people themselves, but also outside people are participating. Members of the community and its current leaders feel that townspeople are changing their perspectives regarding the Muisca revival efforts. Earlier, the sacred ceremonies and the use of traditional plants were viewed suspiciously as forms of drug abuse and addiction. Now, urban non-indigenous people come to Sesquilé seeking healing, an alternative vision of the world and something to give their lives meaning. Some have enthusiastically participated and worked with the community.

Former members now want to return to work in the movement, and new people want to become members of the community revitalization process. The leadership will soon need to think about and establish participation guidelines and protocols. This poses new challenges for the Directive Council and the community as a whole. But, as Sra. Rosa de Mamanché, Carlos Mamanché's mother, put it: 'With Carlos we learned who we are, we know where we are going and what we want.'

Tools for Biocultural Diversity Conservation: Community Mapping of Indigenous Peoples' Traditional Lands in Venezuela

Project Contributor: Stanford Zent

Figure 4.20 *Hotï people drying cane for blowguns*

Credit: Stanford Zent

In 1999, the national constitution of Venezuela gave explicit recognition to the land rights and cultural rights of the country's indigenous peoples. Following passage of the new constitution and subsequent demarcation laws, several indigenous groups began taking the initiative to carry out the demarcation of their lands on their own rather than wait for the government to do it for them. The project 'Ethnocartography and Self-Demarcation of Indigenous Peoples' Lands in Venezuela as Tools for Biocultural Diversity Conservation' (35) is a collaboration between researchers at the Instituto Venezolano de Investigaciones Científicas (IVIC) and two indigenous communities in Venezuela – the Hotï of San José de Kayamá, Caño Iguana, and surrounding regions, and the Eñepa of San José de Kayamá. These two small-scale, culturally unique indigenous groups, whose lifestyles and resource use practices are compatible with environmental conservation, are currently faced with strong pressures for social, techno-economic and ideological change (see Plate 12).

The project is an active collaboration between local community members – who are the principal data collectors and processors – and scientists, who act as advisers and assist in data analysis and document preparation. The project supports the indigenous groups in efforts to secure legal ownership and title to the land they occupy in a tropical forest region rich in biodiversity. This goal contributes to conservation of both biological and cultural diversity, as well as the crucial relationships between them, by seeking to obtain exclusive rights to land occupation and use for the Hotï and Eñepa, and by attempting to achieve land and resource security for these two groups. Community members are being trained in community-based mapping and documentation to produce the necessary cartographic, demographic and cultural-historical documents to support their land claims. Members of local Hotï and Eñepa communities are the principal data collectors and processors and work alongside scientists who serve as advisers, trainers and auditors of the data collection process and provide assistance in data analysis and document preparation. The documents being prepared include maps of current and ancestral territories, community censuses, residential histories, oral histories of human–ecosystem interactions, information about land use patterns and resource management practices, cultural norms and notions about territoriality, property, local and ethnic group membership, environmental ethics and eco-cosmovision and ethnogeographical concepts. The project has also led to greater conscious awareness of the value of traditional knowledge of the environment for the current and future lives of the people. Changes in language are also being documented in order to examine links between language, traditional knowledge and environmental change. The final maps, project reports and other supporting materials were completed in August 2006 and provided to the appropriate National and Regional Demarcation Commissions. This marks the first step in the formal application for land rights recognition. Future plans for the project include adapting the database, maps and reports into educational materials for use by the Hotï and Eñepa communities.

The main challenges faced by the project were mostly technical and logistical. Rather than having outside researchers come in and do the job, the project design emphasized active local participation in different phases of the project. This meant that local capacity building and transfer of information and technology were primary goals along with map making. The idea was not simply to make a map but also to help local people become map makers and map users. In that sense, the project plan called for local people to do most of the basic data collection and processing work. The data collection involved having small work teams traverse different sectors of their territory and record the geographic coordinates of the places that were significant for them, along with the place names and a brief description of the cultural significance in their own words. The local people had to be taught how to read and record data from GPS machines. Since they had no previous experience with this technology, a few weeks of training and supervised repetitive use were spent with each group.

After the field data were recorded on the data sheets, it had to be entered into a computerized database. The communities had no prior hands-on experience with computers and had to be taught computer literacy from scratch. This required a sustained effort over a period of many months. Computer training sessions were held both at IVIC and the home communities, with candidates initially selected by the communities themselves. A typical training session would last two weeks, although frequently the training steps had to be repeated several times for the trainees to become familiar with the tasks. The researchers were also required to review the recorded results on a periodic basis to correct data entry errors. Because of the great distance between the participating communities and IVIC, as well as the high cost of travel between the two places, trips of researchers to the communities or of local collaborators to IVIC could only be accomplished a few times a year. Furthermore, the local participants frequently had other activities and responsibilities (for example, some were school teachers, others were household heads with a family to feed) and were only able to dedicate themselves to the project in their free time. The overall result was that the work was carried out on a piecemeal, rather than a continuous, basis.

Another reason for the slow pace of work was the sheer scale of the land area that was mapped. In the case of the Jotï this was about 5000km^2 and in the case of the Eñepa it was about 3000km^2. The GPS work teams had to undertake long treks and camping trips lasting from several days to more than a couple of months. The overall result was that the field data collection and processing phase of the project lasted for more than three years.

Another challenge has been government inaction. The principal overall goal of the project was to produce the documentation necessary for the two groups to win legal title over their lands. Therefore, the ultimate measure of success of the project must be the official grant of land title to the communities. As yet, this goal remains to be realized. Even though all of the legal documents were completed and handed in to the proper and officially designated authorities in 2006, no firm or concrete decision has been made so far regarding the land title claimed by the Jotï or Eñepa or any other indigenous group that has made the same kind of application. Some observers have questioned the willingness of the government to honour in practice the commitment it has made on paper to claim the land rights of its indigenous peoples. There are several reasons for doubting the government's willingness to live up to its commitment. First, since the new Indian land law was passed, the only land titles that have been given out are small, community-based, usufruct type titles that are essentially equivalent in scope and property rights to the provisional land grants that were made in previous government administrations under the 1960s agrarian reform law. Second, the government agencies in charge of administrating the process have repeatedly changed the requirements for application. Moreover, some groups that have submitted applications have been told that the documents had been lost and had to be resubmitted. Third, even though the law states that the government will fund the process, so far funds have been hard to obtain through official channels. As a consequence, the indigenous groups that have undertaken their own demarcation projects and managed to amass the documentation that is required are still waiting for a concrete decision from the government.

Protecting Territories and Biodiversity: Indigenous Capacity Building in Ecuador

Project Contributor: João de Queiroz

Figure 4.21 *A high level of community participation and capacity building in the development of resource management plans helps foster biodiversity conservation on indigenous lands in Ecuador*

Credit: João de Queiroz

Ecuador's Yasuni National Park comprises almost one million hectares of exceptional biological diversity and includes species such as the giant river otter, jaguar, harpy eagle and 62 species of snake. Adjacent to the park is the Huaorani Indigenous Territory, with 600,000ha that have been declared a UNESCO Biosphere Reserve. Indigenous peoples living in this region are Ecuador's poorest and most vulnerable, yet they control huge areas of primary tropical forest. They are threatened by lack of security and are confronting serious and growing problems with illegal encroachment and colonization, including land clearing for agricultural purposes, continued petroleum production as well as illegal logging, mining and tourism companies, all of which is causing a rapid transformation of indigenous cultures. According to the Ecuadorian constitution and legal framework, indigenous communities must gain legal rights to their ancestral territories before they can acquire the constitutional right to be consulted prior to the initiation of extractive activities within their territories. This process involves a number of legal steps, such as demonstrating that they have occupied the territory for a certain period of time, as well as establishing their boundaries.

The project 'Conservation in Managed Indigenous Areas (Conservación en Áreas Indígenas Manejadas CAIMAN)' (8), was funded by USAID (www.usaid.gov/locations/latin_america_caribbean/environment/country/ecuador.html) and implemented by Chemonics International Inc. in 2002–2007, in consultation with indigenous peoples' organizations (IPOs), primarily representatives of indigenous federations. The project focused mainly on supporting the Awa, Cofan and Huaorani indigenous groups and their respective federations, although it also provided some support to Chachi, Siona, Achuar, Kichua and Secoya populations. Work plans were developed through a combination of workshops with IPOs and consultations with organizations that have worked with these groups for many years. The project's goals were to foster biodiversity conservation by helping secure indigenous legal rights on ancestral lands; strengthen cultural identity and key cultural elements such as language and traditional medicine; and promote income-generating activities that are compatible with the local indigenous communities' socio-cultural and environmental setting, and are both ecologically and economically sustainable (for example, ecotourism – which is not feasible without a healthy ecosystem – and the production of handicrafts). In addition, by ensuring that indigenous peoples were fully integrated into the development and implementation of work plans, the project enhanced capacity for biodiversity conservation within indigenous federations.

Training of forest guards and implementation of regular patrolling of 300,000ha of Cofan indigenous territory have resulted in the eradication of coca fields, the reduction of illegal fishing and hunting, and exposure of illegal mining. CAIMAN helped delimit and mitigate conflicts along 102km of the Huaorani territory and along 105km of territorial boundaries in the areas of Cuyabeno and Cofan-Bermejo, two highly threatened and biodiverse regions of the Amazon Basin. In addition, with CAIMAN support, Kichua communities obtained management rights over 100,000ha in the Yasuni National Park. Three Chachi communities in the buffer zone of the Cotacachi-Cayapas National Park now receive payments for the provision of environmental services and Chachi guards are responsible for ensuring that the environmental integrity of their territory is maintained. Recently the Minister of Environment issued a decree establishing the Awa's full legal title to 100,000ha located within the National Forestry Patrimony, a significant precedent for human rights and conservation.

Promoting Cultural and Biological Diversity: An Educational Programme for Rural Communities in Peru

Project Contributors: Jorge Ishizawa with Grimaldo Rengifo

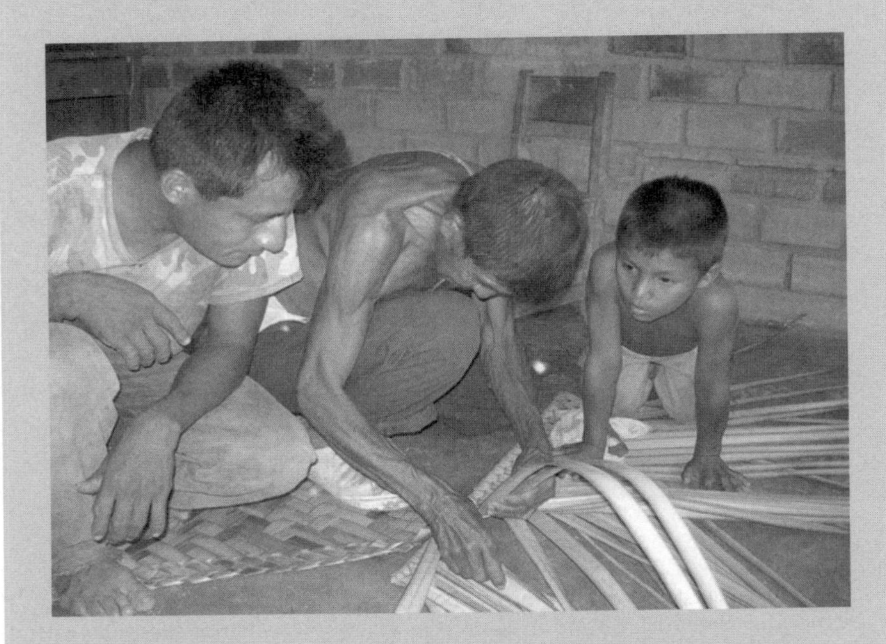

Figure 4.22 *Teaching the children in the Upper Amazon region of Peru*

Credit: Jorge Ishizawa

The Peruvian Andes are recognized as a major site of biological diversity in the world. The Andes have 82 of the planet's 103 life zones, that is, 80 per cent of the ecoclimatic zones existing on the planet (Valladolid, 1998). These range from the coastal desert area to the arid western slopes, to the inter-Andean valleys, to the mountains. As well, the central Andes are one of the eight centres of origin of agriculture, the domestication of plants in this region dating back at least 8000 years (National Research Council, 1989). The region also exhibits the highest inter- and intra-specific agrobiodiversity in the world. This diversity is found in the peasants' *chacras* or cultivated fields, and is due to the care, protection, affection and respect with which peasants nurture their plants. Among traditional societies in the region, an attitude of respect is central to life and is essential for nurturing diversity, both biological and cultural. Respect is expressed in relations between Andean communities and their deities, between human beings and natural entities, and between humans. Andean peasant culture and agriculture are inextricably linked. One cannot be understood without understanding the other (see Plates 7 and 8).

At present, the major characteristic of Andean rural life is the peasant community and small farmer production. According to the 1994 agricultural census, 84 per cent of the 1,764,666 agricultural units were peasant farms of less than 10ha. No other economic sector in Peru incorporates as large a population – over 7 million people. However, peasant communities own only 10 per cent of the total agricultural land. As of 1998, peasant communities numbered almost 7000 and were located in diverse ecosystems of the coastal area, the highlands and the Amazon region. Andean Amazonian peasant agriculture is based on local practices and inputs, and still produces a major part of the fresh food that reaches urban markets. Over time, however, there has been a general loss of respect among people in the region, and this has come to constitute a threat to biodiversity conservation.

In 1969, General Velasco's government decreed an agrarian reform, one of the most radical changes in the rural property regime of the Peruvian Andes. This reform had the explicit aim of promoting industrial development through rural modernization. Throughout the Andes, large haciendas were transformed into cooperatives and associative firms, owned by the former hacienda workers or by communities. Eventually these firms went bankrupt and the lands were distributed to individuals or communities. The results of four decades of 'development programmes' were already evident by the end of the 1980s. Peru had not only become more dependent on import substitution as a result of the industrialization process, but the agricultural indices for production and productivity had also decreased. The country had joined the roster of net food importing countries in the world. Development had not fulfilled its promise, and development had been predicated on the eradication of the native cultures as the price to be paid for progress.

The 'Andean Project for Peasant Technologies' (Proyecto Andino para las Tecnologías Campesinas, PRATEC) (15) is a Peruvian NGO founded in 1988 and devoted to the recovery and valorization of traditional agricultural practices and associated knowledge. PRATEC participates in the efforts of Andean Amazonian peasant communities to counter the socially and ecologically destructive effects of industrial agriculture and governmental agrarian policies. By using local knowledge and the practice of traditional 'ritual agriculture' and through adopting a non-dualistic, eco-centric worldview, PRATEC supports the resurgence of local approaches to agriculture, which it sees as radically opposed to Western industrial agriculture. The Andean peasant practice of ritual agriculture embraces kinship-oriented visions of the land and encourages empathetic actions that illustrate respect for all living entities of the biosphere. Agricultural activities include ritual actions, utterances and offerings that express both a deep respect for *Pachamama* (Mother Earth) and communitarian aspects that characterize the worldview of the Andean people (http://fore.research.yale.edu/religion/indigenous/projects/pratec.html).

During the decade of the 1990s, PRATEC's institutional efforts were devoted to the documentation of peasant agricultural practices and training through an annual course on Andean peasant agriculture. Around 140 university teachers and technical personnel of rural development projects were trained. The unexpected outcome was the formation of community-based organizations, called 'Nuclei for

Andean Cultural Affirmation' (NACAs), small NGOs that presently support rural communities in six regions of Peru. The NACAs work with families, who traditionally nurture biodiversity in their *chacras* to help them remember the ways in which their ancestors learned respect for the land (see Plate 7). An initial six-year programme with six NACAs made clear that, beyond increases in production and productivity, *campesinos* see biodiversity conservation as intimately related to the maintenance of a worldview, or cosmovision, based on respect and affection. Agricultural practices in the Andes, including soil preparation, seed diversification, sowing, harvesting, storage and food preparation, can only be understood in the context of such cosmovision. The idea of the annual course was to train people to understand and interpret this cosmovision. The goal of PRATEC's programmes with the NACAs has been to recover the respect for biodiversity among all members of the local communities.

An in-situ conservation project carried out in 2001–2005 aimed to stop the genetic erosion in the diversity of native cultivated plants and their wild relatives in the central Andes. The extraordinary inter- and intra-specific diversity of plants and animals that has been nurtured for millennia by *campesino* communities was threatened by the modernist spread of monoculture. Consequently, the project's overall objective was to conserve agrobiodiversity in the *chacras* of *campesinos* in 52 locations in Peru. The project addressed six areas of intervention:

1. the *chacra* and its surrounding space;
2. the social organization of in-situ conservation;
3. awareness of the importance of maintaining the diversity of native cultivated plants and their wild relatives;
4. policies and legislation to promote in-situ conservation;
5. market development for agrobiodiversity;
6. an information system for monitoring agrobiodiversity.

The project found that agrobiodiversity is the result of Andean Amazonian agricultural practices. Here, as in other original agricultural areas, making *chacra* is not a 'way of making a living' but a way of life. *Campesino* Don Humberto Valera, from the Upper Amazon region of San Martín, clearly expressed this in talking about making *chacra*: 'It seems that we will never finish harvesting this *porotal* (bean *chacra*). You produce a lot when you know how to endear yourself to the *chacra*. Several different varieties appear, some others return' (PRATEC, 1998, p4).

Don Cristóbal Ramos Rosa from the community of Calacoto, Corcori in Yunguyo, Puno understands in-situ conservation in the following way: 'For the *paqalqus* (groups of families) who nurture diversity, making *chacra* is a permanent concern. It has always been this way and will continue being so. To adapt to difficult circumstances we make offerings to all *uywiris* [nurturing deities]. We converse with the *pacha* [local world, including deities, humans and natural entities] very affectionately and ask our deities to prevent the *ispallas* [ritual name for tuber seeds] leaving us because of our mistreatment. Likewise, we make offerings to the spirits of frost, hail and drought who nurture us. Our authorities [past and present] are the ones who are concerned with making us converse, and the Andean priests convoke the human communities to the top of the mountains to ritually ask our seeds, *Pachamama*, mountain deities, for forgiveness... When we have attained peace among ourselves and with the whole *pacha*, the *chacras* become vigorous and happy. We return to our places to continue nurturing our *chacras*, respecting and obeying our authorities. This is the way we do it, always with affection and with all our hearts, with rituals, festivals' (PRATEC, 2006, pp37–38).

Centering on the recovery of respect in the communities involved in the in-situ project, the NACAs endeavoured to recover and/or strengthen the traditional authorities of the *chacra* and the *sallqa* (the wild). This was attained through the strengthening and/or revival of rituals and festivals in the agricultural cycle. Visits between communities for seed and knowledge exchange were also instrumental in the mutual learning that led to the recovery of community memory about how their ancestors lived in sufficiency based on diversity (http://video.google.de/videoplay?docid=2648819621498941167&q=source:013333240103878510926&hl=es). The project was successful, especially in showing that vigorous practices of in-situ conservation were still widespread in many places in the Andes and the Upper Amazon region, and even if the spectacular increases in agrobiodiversity in the participating communities may not prove sustainable without external intervention in the long run, the threat of genetic erosion does not appear to be imminent. A more immediate result has been the growing national awareness and pride in being a mega-centre of biodiversity, which is expressed in the international recognition of the excellence of Peruvian cuisine based on the diversity of native plants.

During the period 2002–2007, PRATEC conducted a programme called 'Children and Biodiversity', coordinating the fieldwork of six NACAs located in the Andean highlands. The programme had an important educational component that sought to incorporate local knowledge into the school curriculum and to involve parents in school activities. The focus of the programme was to explore the possibility of the community nurturing its school. It also aimed at restoring the autonomy and authority granted to children in the traditional system of governance, as in the past children were able to exercise control within the community, for instance taking care that animals did not enter the *chacras* and sanctioning those who let their animals trample their neighbours' crops (see Plate 8).

These initial aims were in accordance with the traditional authorities in the communities, who had been unanimously pointing to 'loss of respect' as the main obstacle for community well-being. The educational system was identified as a major threat to the conservation of the diversity of native plants because respect and affection among entities of the *pacha* had been eroded by the imposition of a system that disparages local traditional worldviews. The signs were clear: 'Children no longer greet their elders'. This was after 50 years since these same communities had demanded that the educational system help transform their children and equip them with skills so they could migrate to the cities and to a life of 'progress'. Children were to be transformed so they were prepared to live in a future of 'progress' instead of a present that was regarded as backward and inferior.

In discussion with parents in 2004, it was made clear what the traditional authorities wanted from the school. This was expressed as *Iskay Yachay* in Quechua and *Paya Yatiwi* in Aymara. They both translate into 'two kinds of knowledge' – their own and the school's. The documentation of the local knowledge of conservation practices included in local traditions became the basis of the school curriculum. The project adopted an intercultural approach allowing the coexistence of diverse 'educational cultures', that is, modes of intergenerational knowledge transmission of a given community (Rengifo, 2005). This concept is particularly useful in order to go beyond the dualism between home-based vs. school-based local/indigenous knowledge transmission. The project strategy included the training of rural teachers as cultural mediators, capable of integrating local knowledge into the school curriculum, as well as the consolidation of orality as a basis for literacy. The central finding of the Children and Biodiversity Project is that *Paya Yatiwi/ Iskay Yachay* has three interrelated components: (1) the recovery of respect in the community (towards their deities and nature and among the community members themselves); (2) learning to read and write while respecting and valuing the local oral traditions; and (3) teaching the skills to allow people to live a good life (http:// video.google.de/videoplay?docid=889973005344471523&hl=es#, http://video. google.de/videoplay?docid=-8141771985988154873&q=Pratec&ei=JFCPSL2iDYe MqQLxvPSADA&hl=es).

The Children and Biodiversity project has been successful in clarifying the challenges that must be faced by intercultural education. The incorporation of local knowledge into the school curriculum and the adoption of the local agricultural calendar have become a national policy. The three components identified in the case of rural education have inspired other institutions, especially in the southern Andes, to initiate training programmes for rural teachers. Networks of rural teachers have been formed in the localities where the programme was active and provide the surest guarantee of the sustainability of the programme results. This process of cultural 'regeneration' takes time since the communities themselves must find them relevant to their own life world. Meanwhile, the training of educators continues, as this process requires not only a new attitude and conceptual framework, but also an alternative to training by mainstream 'rural development experts'. Since 2002, PRATEC has offered three versions of a two-year Masters' programme on Biodiversity and Andean Amazonian peasant agriculture, in cooperation with the Universidad Nacional Agraria de la Selva (UNAS), a state university based in Tingo María, in the upper Amazon region. Almost 60 university graduates have participated in the programme. Under the same agreement, versions of an annual diploma postgraduate course on Intercultural Education and Sustainable Development are being offered to rural teachers and rural development workers of the Andean region including Ecuador, Peru, Bolivia and northern Argentina.

A 'Life Plan' for the Park: Culturally Appropriate Management in Brazil's Xingu Indigenous Park

Project Contributor: Darron Collins

Figure 4.23 *Mapping traditional territories in the Xingu Indigenous Park*

Source: Amazon Conservation Team

The concept of 'National Park' in Brazil incorporates the dual objectives of protecting the environment and the indigenous populations living within its boundaries. Parks are administered by the National Indian Foundation (known in Brazil as FUNAI) and the Brazilian government's environmental agency. The Xingu Indigenous Park, a 6.5 million acre area of tropical forests and savannah in central Brazil, is inhabited by the Xingu peoples, a coalition of 14 indigenous groups totalling over 4000 individuals. The park was created in 1961 by the government of Brazil to mitigate the degree to which its isolated communities would be disturbed or destroyed by colonizing forces. In the 1980s, however, hunters and fishers started invading the territory of the Xingu Indigenous Park. A booming agricultural industry and the encroachment of cattle ranching in the region, along with a lack of federal resources to adequately enforce the park's boundaries, created a situation that put mounting pressures on the integrity of the reserve and the communities within it. By the end of the 1990s, forest fires on cattle ranches located to the northeast of the park and the advance of forestry operations to the west also threatened to affect the park. Further, the occupation of the area around the park began to pollute the headwaters of the rivers that supply water to the park. Due to these pressures, there has been an ever-increasing perception among the indigenous inhabitants of the park that they are in an uncomfortable 'embrace', surrounded by a process of occupation, and that the park is a shrinking 'island' of forest in the midst of pasture and intensive agriculture. With a growth rate of around 3 per cent per year, the population has nowhere to expand. Therefore, life in the villages follows a progressively sedentary pattern, in contrast to traditional semi-nomadism. Activities that surround the park are preventing the flow of sources of animal protein (game animals) into the park, so the availability of natural resources is becoming increasingly scarce.

Another challenge to the people in the Xingu Indigenous Park is rooted in the history of the formation of an internally diversified Indigenous Area – both from a socio-cultural and an ecological perspective. Several indigenous societies have had to coexist in a situation of geographical confinement. Further, new indigenous organizations (largely the Indigenous Land Association of the Xingu, known in Portuguese as ATIX) have been established as a means for dialogue with the national society and to encourage projects in education, economic alternatives and protection of the territory. These organizations are using an administrative structure that does not exist in the traditional political structures of the indigenous societies, and that presupposes command of the Portuguese language, basic mathematics, legislation and inter-institutional relations. Younger individuals are the ones who dominate the new knowledge that is indispensable at this interface. This generates conflict with traditional village politics, which is generally controlled by elders. Thus, an indigenous association does not always succeed in reconciling the traditional politics with the political administration of the national society.

Since 1996, the Xingu peoples have been working together with the Amazon Conservation Team (ACT) toward the goals of abating illegal incursions into the park and establishing a culturally appropriate management scheme, called a 'life plan', for the park and its inhabitants. The project, 'Territorial Management in Brazil's Xingu Indigenous Park' (24), developed maps of traditional territories, sacred sites, fishing and hunting locales, and other salient features of the landscape to drive the conservation of biodiversity in the park. Initial activities involved biocultural mapping for Kamayurá indigenous ancestral lands, upon the direct invitation of the tribe. ACT equipped the indigenous researchers with handheld GPS units and provided training in ethnographic map composition, while Western-trained cartographers assisted with the technical map assembly. In 2002, ACT and its indigenous partners completed maps of the Kamayurá and Yawalapiti areas of the Xingu Indigenous Park, covering 1,250,000 acres. In the process, ACT worked in collaboration with FUNAI and with the Pilot Program to Preserve the Brazilian Rainforest. The maps were released in a three-day ceremony in the Xingu. Since then, ACT and its indigenous partners completed the collaborative mapping process for the entire Xingu Indigenous Park, an area of over 7 million acres of savannah and lowland tropical rainforest.

The process of ethnographic mapping in the Indigenous Park and the construction of management plans based on these maps have brought a significant degree of focus and attention on the knowledge of older generations. In addition, the incorporation of technological devices like GPS units and mapping equipment has intrigued the younger generation. Thus, these various elements of the project have united people across generations and inspired new admiration and respect for elders and their knowledge. This mapping project also brought together the 14 tribes of the Xingu, representing the first time they ever worked together to complete a single project. 'This is truly the first effort in the history of our territories that has united our 14 tribes toward a common end, and these are the first maps to be published in our native languages', commented Tunuly, a Yawalapiti tribe member.

The project has found that processes that occur outside the park directly affect what happens inside the park. Therefore, the sustainability of the park depends on developing ways of doing politics outside the park, identifying possible allies and seeking to sensitize the relevant public agencies and the public in general to what is happening in the region of the Xingu, having to do with defending both the indigenous population and the biodiversity of the Amazon forest. Also, importantly, in 2007 ACT concluded that the leading indigenous organizations of the Xingu Indigenous Park were able to manage their own land and cultural conservation efforts, and passed on the assets of its regional field office in Canarana to the associations of the Waurá and Ikpeng. ACT now serves in an advisory capacity for indigenous associations of the Xingu seeking assistance in conservation and sustainable development planning and implementation.

Protection of an Indigenous Reserve: the Ka'apor People of Amazonian Brazil

Project Contributor: William Balée

The Ka'apor emerged as a people with a distinctive identity about 300 years ago, probably between the Tocantins and Xingu Rivers in the Amazon Basin. They later engaged in a long and slow migration that took them into Maranhão State, in eastern Amazonian Brazil, by the 1870s. One hundred years later, in 1978, the Alto Turiaçu Indigenous Reserve (called Terra Indigena Alto Turiaçu today) was demarcated by Brazil's FUNAI. The reserve covers about 5300km^2 of high Amazonian forest and is inhabited by all remaining Ka'apor as well as by some Guajá, Tembé and Timbira people. The Ka'apor, like many other settled Amazonian groups, are a horticultural people whose staple is bitter manioc. They grow about 50 domesticated plants, which are used for food, seasoning, medicine, fibre, tools and weapons. In addition, they hunt game and gather fruit in the dense forests and fish in the tiny creeks of the reserve. Since the late 1980s, as much as a third of the reserve has been illegally deforested and converted to towns, rice fields and cattle pastures by landless peasants, cattle ranchers, loggers and local politicians. The present situation is marked by tension and escalating violence. Raids on indigenous villages by squatters and loggers and counter-raids by native people on squatters' and loggers' camps inside the reservation have occurred since 1993 with at least two fatal casualties.

In 2003, the Urubu-Ka'apor, with the support of the World Wildlife Fund, established a non-profit corporation, called the Associação do Povo Indígena Ka'apor do Rio Gurupi; Associação Ka'apor; Associação do Povo Indígena Ka'apor Ta' Hury, designed to support activities related to health, education and sustainable management of resources, culture and environmental protection of pre-Amazonian forests in western Maranhão state. The project 'Jande Myra Ta Ka'a Rupi Ha (Our Trees of the High Forest): Ka'apor Ethnodendrology' (3) works in tandem with the Ka'apor non-profit corporation and is sponsored by the Museu do Índio, Rio de Janeiro, Brazil. The project involves dissemination and protection of the history, customs, arts and traditional cultural practices as well as the language of the Ka'apor people as these relate to the forest. In particular, it seeks to aid the Ka'apor in preserving knowledge of the trees found in the Ka'apor habitat and in protecting that knowledge, along with their arboreal legacy itself, from usurpation by external commercial logging interests. Extension work on the project, involving the training of indigenous fellows in tree photography and knowledge, began in August 2009. There is a feature exhibit of the Ka'apor in the National Museum of the American Indian in Washington, DC, which showcases the Ka'apor's concerns over their land, trees, culture and languages. The new project with the Museu do Índio also envisions a feature exhibit on Ka'apor trees as well as many other products, to be developed among the Ka'apor people themselves.

Training Indigenous Agro-Forestry Agents in Acre, Brazil: Indigenous and Modern Technologies for Sustainability

Project Contributors: Giulia Pedone and Renato Gavazzi

Figure 4.24 *The Kaxinawá people of Acre, Brazil*

Credit: CPI/Ac archives

The Amazon region has largely been perceived as a boundless territory with unlimited resources to exploit. Due to its low population density, it has been viewed as an 'empty space' to be colonized and to be integrated into the national economic landscape, and thus as a key to Brazil's progress as a 'modern' nation. During the 1960s and 1970s, the military government promoted a media campaign to encourage private owners to invest in the Amazon region – the national slogan was 'a land without men, for men without land'. This resulted in marginalized farmers from the poorest regions of Brazil moving into the Amazon rainforest in quest of a better life. Over the past 35 years, the forests of the state of Acre in the western Brazilian Amazon have also been adversely affected by large-scale Brazilian economic interests, backed by financial resources obtained from credit institutions and by Brazilian government incentives for the establishment of large cattle ranches, the exploitation of hardwood and agricultural activity. These incentives have led to considerable concentrations of private property, and serious conflicts have resulted from land takeovers, which have provoked confrontations between the 'new owners of Acre' and the local indigenous populations and rubber extractors. This has led to a progressive loss of biodiversity and a scarcity of traditional sources of protein, which is evident in the increasingly deficient diet of the indigenous peoples in these areas.

After the process of legal allotment and demarcation of indigenous territories took place in the 1970s, three new professions developed among the Amazonian indigenous peoples in order to assist the local indigenous groups in managing their own territories: bilingual teachers, health workers and Indigenous Agro-Forestry Agents (IAFAs). In the project 'Training Program of Indigenous Agro-Forestry Agents of Acre' (28), indigenous peoples from seven different indigenous nations of Acre received training on the theory and practice of natural resource management, with the support and guidance of the NGOs Commisão Pró-Indio do Acre (CPI/Ac) and Associação do Movimento dos Agentes Agroflorestais Indígenas do Acre (AMAAI-Ac), in response to political demands from regional indigenous populations. The main issue for indigenous peoples returning to their native lands is how to be economically active, culturally relevant and ecologically sustainable on their lands after being employed as labourers on rubber plantations and in agricultural operations. The project involves training related to agroforestry systems, the improvement of degraded areas, management of palm plantations and techniques of livestock management. Awareness of environmental legislation and domestic policies related to demarcation of indigenous territories is also a part of the training programme.

The programme operates on the belief that blending indigenous and modern technologies enhances the ecological sustainability of the indigenous territories. The Indigenous Agro-Forestry Agents act as 'environmental educators' who work to revitalize the indigenous traditional ecological knowledge, to preserve and strengthen cultural diversity and establish a sense of identity and social cohesion. IAFAs receive a bilingual and intercultural education in their own native languages as well as Portuguese. The main results include enabling indigenous peoples in the region to manage and conserve their demarcated territories by instilling the capacity to develop alternatives for the sustainable management of their environment. Project success is shown by the number of trained IAFAs, which increased from 15 in 1996 to a current level of 126, originating from 11 indigenous areas. In addition, the IAFA training has become a working model that has been replicated in other Brazilian states (see Plate 6).

Future challenges for the indigenous peoples of Acre include finding adequate and endogenous solutions to manage their territories in harmony with their cosmovision and their perspectives on the life they want to live; and building a relationship with the national society based on equal interchange and mutual collaboration. In the words of one Indigenous Agro-Forestry Agent: 'We are indigenous environmental educators... The forests are the greatest wealth that our land, our state, our country has. We hold meetings, we discuss, we teach and we guide our relatives on environmental management. We are concerned with the destruction of the planet and we want our forests to stay standing, giving us the strength that we need during our short lives.'

Global

The Language of the Environment: A Comparative Environmental Thesaurus

Project Contributor: Fulvio Mazzocchi

The 'Environmental Applications Reference Thesaurus (EARTh)' project (11), carried out by the Institute for Atmospheric Pollution at the National Research Council in Italy, is developing an advanced tool to be used for environmental information management and environmental policy and research. The project's aim is increasing awareness among policy makers of the complexity of the environmental domain and of the cultural dimension of environmental knowledge. Thesauri are controlled vocabularies designed to allow for effective indexing, classification, cataloguing and retrieval of information. They consist of a network of semantic relationships, by means of which a representation of the meaning of each thesaurus term as well as of the conceptual structure of a knowledge domain is provided. Thesauri can be regarded as 'semantic road maps' for information indexers and searchers and for anybody else interested in a systematic grasp of a given field. Existing terminological or knowledge organization systems at the international level do not provide an adequate and updated account of the environmental domain. To meet the present needs of environmental information management, more refined semantic structures are required, in the form of thesauri.

The EARTh thesaurus focuses on a broad spectrum of environmental terminology, but it contains a conceptual and terminological section specifically on biodiversity and will provide a foundation for information management on knowledge related to biodiversity. Sources for the thesaurus include international documents such as the terminological bulletin used at the Environment and Development conference in Rio de Janeiro in 1992. The thesaurus is strongly oriented toward the cultural dimension of environmental knowledge and knowledge organization – an important goal in times when diverse cultures with distinct visions of the world need to work together to address environmental problems. The project recognizes that traditional knowledge classification systems and environmental terminologies encapsulate traditional worldviews and reflect an indigenous cognitive structuring of reality. In order to better represent indigenous and traditional cultures within the global context, these systems and terminologies are planned to be included within the thesaurus in the form of special annexes. For this purpose, partnerships for documenting indigenous and traditional terminologies and classification systems will be established, subject to availability of a technological infrastructure able to handle different languages of the world.

With limited economic and human resources, EARTh is still in a phase of implementation, involving the core vocabulary and semantic structure. One of the main challenges about this work is related to the delimitation of the environmental field – since it also implies policy, economic and social aspects. Another challenge is the fact that environmental concepts and terms could be interpreted differently according to different disciplinary and cultural views, implying that multiple semantic structures are to be handled within the same system.

Strengthening Culture and Conservation Through Intangible Heritage and Performing Arts: The 'Dance for the Earth and for Her Peoples' Initiative

Project Contributor: Robert Wild

Figure 4.25 *A dance of the Bambuti community of Semliki Forest, Western Uganda*

Credit: Robert Wild

The concept for the 'Dance for the Earth and for Her Peoples' initiative (10) originated at the 2003 World Parks Congress and has been taken forward by the IUCN World Commission on Protected Areas (WCPA) and the Commission on Environmental, Economic and Social Policy (CEESP) through the Theme on Indigenous and Local Communities, Equity, and Protected Areas (TILCEPA) and the IUCN Task Force on the Cultural and Spiritual Values of Protected Areas (CSVPA). The objective of this initiative is to explore the role of community performing arts in strengthening the conservation of biocultural diversity, especially in Indigenous and Community Conserved Areas (ICCAs).

Intangible heritage and the performing arts are a strong force in social cohesion and the intergenerational transmission of cultural knowledge. Many traditional dances, for example, have strong links to nature and the landscape, as they borrow movements from animals, express seasonal and annual cycles, and act out stories related to nature. The Dance for the Earth initiative aims to test the use of the performing arts as a tool for promoting the conservation of biocultural diversity at a representative selection of protected areas around the globe; build a network of institutions, organizations and individuals interested in the initiative; develop and fund a number of dance-related projects at protected areas; collect, record and conserve 'Earth dances' from different cultures around the globe. Through dance and drama, communities strengthen the links between conservation of nature and the maintenance of culture, and community groups tell their stories and celebrate their efforts to conserve their traditional lands and enhance sustainable livelihoods. The initiative is spearheaded by a diverse international network, predominantly made up of conservation professionals who have direct contact and work with community groups in different parts of the world. As such, it has been developed in a participatory way, and a number of local communities have enthusiastically taken up the idea.

Worlds of Difference: Local Culture in a Global Age

Project Contributor: Jonathan Miller

Figure 4.26 *Peruvian biologist María Scurrah learning the names of traditional potato varieties from a farmer in Quilcas, Junín Department, Peru*

Credit: Jonathan Miller

Homeland Productions (http://homelands.org) is an independent, non-profit journalism cooperative in Tucson, Arizona, US, specializing in radio documentaries. Its mission is to illuminate complex issues through compelling broadcasts, articles, books and educational forums, and to foster freedom of expression and creative risk through the media arts. Homeland reaches tens of millions of radio listeners through its features for mainstream news and public affairs programmes in the US. The 'Worlds of Difference' radio series (30) was produced by Homeland for national broadcast on public radio stations in the US. It used both radio and the internet to generate awareness of how people with strong local traditions are responding to the pressures and opportunities of rapid cultural change.

The radio series (http://homelands.org/worlds), which ran from 2002 to 2005, produced 40 feature stories from 27 countries and six hour-long radio specials. The specials, organized by theme (economy, language, religion, home, history, future), were distributed by National Public Radio and aired on more than 100 non-commercial radio stations in the US. Teachers in several states report that they have incorporated the website and the audio (available online) into their curricula.

One of the project's documentary features focuses on Andean potato farmers and the pressure on them to convert land and labour into cash, which threatens their vast traditional knowledge of agrobiodiversity, including knowledge of hundreds of varieties of potato. Farmers are compelled to replace potato and other crop varieties with more marketable (or higher-yield) 'improved' varieties, or to concentrate on growing a smaller number of traditional varieties for sale to niche markets. Thanks mostly to the work of NGOs and scientific organizations, there is increasing, although still not widespread, public recognition in the region of the need to conserve agrobiodiversity. Some farmers now compete in fairs for the highest number of varieties of several traditional crops, and trade recipes for 'value-added' products like jams and chips.

Other pieces in the series are concerned with language revitalization efforts that show varying degrees of promise (Welsh, Maori, Occitan, Zapotec, Ladino). A story about the Zápara people of Ecuador documents the case of a community facing seemingly impossible odds. The Zápara, once the most numerous people of Ecuador's Amazon region, had come close to extinction, with only four remaining speakers of their language. Efforts to record and recover the language are underway, but the process is fraught with threats, the most urgent being the presence of oil concessions in their area of the Amazon.

In the words of one producer, the common thread in several of the stories presented in the series is that 'the central drama, as defined by the protagonists themselves, was how to conserve what was unique to them as a people while moving forward economically.' The Worlds of Difference website includes streamable audio, photographs, articles and a sampling of quotes about cultural change. Especially poignant are the words of Octavio Paz, Mexican poet: 'The ideal of a single civilization for everyone implicit in the cult of progress and technique impoverishes and mutilates us. Every view of the world that becomes extinct, every culture that disappears, diminishes a possibility of life.'

5

Cross-cutting Analysis of the Projects

Ellen Woodley

The project descriptions in Chapter 4 illustrate the close interdependence of human cultures and the ecosystems they are a part of. They highlight the diversity of approaches used to integrate biodiversity conservation with support for cultural resilience. Together, these projects provide a rich source of innovative problem-solving to address the global decline of biocultural diversity. In this chapter, we analyse the projects collectively and make generalizations, based on the information obtained through the survey and further exchanges with sourcebook contributors. Our main purpose here is to synthesize this information in order to identify some of the main factors that affect the permanence or loss of biocultural diversity and to identify some of the key features of the projects based on several analytical dimensions.

To simplify the discussion of so many diverse examples, the projects are grouped according to their main conservation 'entry point'. For example, a project's main goal may be to achieve biodiversity conservation (at the genetic, species or ecosystem level), while also seeking to support aspects of culture associated with that biodiversity. Or the entry point may be the affirmation of languages or cultural practices and knowledge, while also building on the association of language and culture with local biodiversity.

It is important to recognize first of all that biocultural diversity can be supported in two main ways. When traditional customs are alive and well and people and their local environment are not threatened, biocultural diversity can be sustained in an implicit and spontaneous way, through the continued unfolding of traditional values, beliefs, knowledge and practices, as well as the sustained use of local languages. In other cases, support for biocultural diversity can be in the form of explicit and conscious efforts at 'revival'. These are attempts to restore culture, language and the environment after they have already eroded, or to sustain them when damage is imminent or has begun.

Most of the projects included in this sourcebook are of the 'revival' kind, with a focus on conservation and restoration efforts in the face of threats to biodiversity, cultural integrity and language maintenance. However, there are some projects that fall into the 'traditional' category. For instance, the project on conserving sacred sites in Ethiopia (43) documents the continuation of traditional practices in order to convince the government of the biodiversity conservation value of these practices. Another example is the promotion of traditional medicine in Uganda (19), which acts to strengthen the ongoing activities of healers and their relationship to biodiversity. Similarly, the Gwich'in place names project (14) in northern Canada relies upon existing traditional knowledge for land use planning in the region. Also, the traditional crop landraces project in Nepal (17) shows that the maintenance of the traditional life cycle and food rituals contributes to ensuring food security through the continued use of traditional landraces in festivals and ceremonies.

The distinction between 'revival' and 'traditional' approaches is important, because the prevalence of revival approaches in our case studies underscores the pervasiveness of threats to biodiversity and cultural diversity worldwide. At the same time, the traditional approaches illustrated here point to areas of some degree of resilience – areas in the world where traditions have been able to continue on some level without imminent threat, or where they show resilience to such threats.

The first section of this chapter examines some of the common causes of biocultural diversity loss, identified from the diverse perspectives of sourcebook contributors. These factors of change manifest themselves at all scales, but have identifiable impacts at the local level. The next section discusses the links the projects make between biodiversity conservation and cultural affirmation. The subsequent section then reviews the collaborative nature of projects: that is, whether indigenous peoples or local communities designed and implemented the project and/or the extent to which they work in close collaboration with outside researchers. In the following section we focus on how the projects are developing or strengthening methods and institutions for the maintenance and revitalization of intergenerational transmission of local knowledge and languages. The final section examines how some of the projects are implementing or contributing to biocultural diversity policy. Analysing projects in this manner forms the basis for our discussion of lessons learned in Chapter 6 – what we learned from these integrated biocultural projects and what lessons conservation projects worldwide may take from these on-the-ground experiences in biocultural diversity conservation.

Causes of the loss of biocultural diversity

According to the information gathered through our survey, people working on the ground on integrated conservation projects attribute the loss of biodiversity and cultural diversity to a variety of forces of an ecological, economic and social nature. The main reasons given for the loss of biodiversity and the loss of cultural practices, languages and traditional knowledge related to biodiversity are listed in Table 5.1, along with the numbers identifying the projects that mentioned each specific factor. Most of the factors mentioned by project contributors are exogenous to the respective project areas, that is, they are due to external forces. As such, many of these factors – particularly those related to environmental degradation, land use conversions, changes in biodiversity, over-exploitation of natural resources, economic development and land and resource tenure security – often lie largely outside the control of the local communities where the project is located. On the other hand, there are several factors over which communities do have some control, although these too are mostly driven by exogenous forces. Many of these fall into the category of 'acculturation and socio-economic change'. A number of projects address these latter factors directly. For example, many projects report an intergenerational breakdown in the transmission of traditional knowledge and the associated loss of languages, knowledge and traditional beliefs related to biodiversity, due to acculturation. These projects strive to change the situation by means that are available to local communities, such as by bringing elders and youth together (such as projects 14 and 29), or enabling the youth to experience a sense of place on the land (such as project 39).

Table 5.1 *Factors affecting biocultural diversity loss*

Reasons for the loss of biodiversity, cultural practices, languages and traditional knowledge	*Project numbers*
1. Environmental degradation, land use conversions, changes in biodiversity and over-exploitation of natural resources	
Habitat loss	6, 9, 23, 43, 44, 45
Soil erosion	6, 44
Decline of water resources	6, 44
Pollution of watercourses	6, 9, 18
Degradation of marine environment	1, 7
Fires	4, 24, 36
Climate change	1, 6, 7
Deforestation	3, 6, 9, 26, 31, 45
Wetland drainage	9
Agro-industry and monocropping; replacement of traditional crops with non-native species	3, 9, 15, 18, 23, 28, 30, 31, 32
Purposeful extermination of species	9
Encroachment of exotic and invasive species	1, 7, 23, 44
Exploitative commercial forestry	4, 9, 28, 29, 36, 40
Over-fishing or destructive fishing methods	9, 12, 27
Over-hunting	9, 18, 23, 24, 26
Over-grazing	4, 6, 43
2. Economic development	
Urbanization	1, 2, 3, 7, 32, 44
Mining	6, 8, 9, 21, 31
Agricultural and grazing encroachment	3, 6, 8, 9, 24, 28, 31
Tourism	6, 8, 36
Natural gas or oil production	5, 8
3. Land and resource tenure security	
Contested sovereignty and land tenure; illegal incursions on indigenous territories	3, 4, 6, 8, 24, 32, 35
Ineffective state governance	4, 13, 15, 20, 24, 36
Expansion of the state	9, 32
Lack of control over local resources	22, 32
Privatization of collective lands	28, 36
Forced evacuation (nuclear testing)	1, 7

Reasons for the loss of biodiversity, cultural practices, languages and traditional knowledge	*Project numbers*
4. Acculturation and socio-economic change	
Lack of intergenerational transmission of knowledge of local biodiversity (changing socio-economic context, lack of communication between elders and youth, disinterested youth, few opportunities for traditional teachings)	2, 6, 9, 12, 14, 15, 16, 22, 23, 31, 32, 35, 37, 39, 40
Loss of languages and erosion of traditional knowledge and practices	2, 6, 14, 16, 31, 33, 35, 40, 42, 44
Loss of traditional beliefs relevant to conservation of biodiversity	9, 15, 17, 19, 32, 33, 43, 44
Breakdown of traditional education systems; formal education systems that discourage or impede teaching of local language, cultural knowledge and worldviews	2, 6, 9, 15, 16, 36, 37, 43, 44
Disconnection from environmental experience or physical disconnection from 'place'	2, 9, 16, 22, 31
Ideology of progress	15, 22
Immigration of non-indigenous/non-local settlers	9, 20, 28
Out-migration from indigenous/local communities	6
Low self-esteem, general social decline due to colonization	2, 13, 16, 19, 31, 36, 38
Missionization and monotheism	23, 35, 36, 40, 43
Civil unrest, war, violence	3, 6, 9, 36
Poor health services; loss of knowledge and availability of traditional medicinal resources	6, 20, 23
Loss of food security; nutrition/diet problems from insufficient food production or diminished availability of traditional foods	2, 6, 20, 28
Incursion of non-native plants, plastic products into local markets, resulting in increased dependency on imports and decreased reliance on home-produced foods and utensils	23, 37
Lack of recognition of the value of traditional knowledge by outsiders and the state, affecting knowledge maintenance	22, 43
Misappropriation of cultural knowledge in the documentation process	5

Many of the factors of biocultural diversity loss listed in Table 5.1 are well known from the abundance of existing studies on the causes of loss of biodiversity and ecosystem health. A handful of these factors comprise what is commonly considered to be the 'big five' sources of pressure on ecosystems and biodiversity (Rapport and Singh, 2006):

1. physical restructuring (modification) of terrestrial and aquatic ecosystems for development and other human uses;
2. discharge of waste residuals (toxic substances and excess nutrients) into the environment;
3. over-harvesting of natural resources from both land and water;
4. purposeful or accidental introduction of invasive alien (non-native) species;
5. extreme natural events such as hurricanes, tsunamis, fires and floods (which are now greatly enhanced by radical human transformation of land, water and climate).

What is significant in the present context is that, by and large, these same major forces of change are also negatively affecting local cultures, and thus cultural diversity globally.

Figure 5.1 *Commercial logging in the Sierra Tarahumara of northern Mexico is a major source of deforestation, soil erosion and loss of water resources, all of which severely affect local communities*

Credit: David J. Rapport

Furthermore, an examination of the causes of diversity loss from a biocultural perspective also brings to the fore socio-economic and cultural pressures (from issues of land and resource tenure to forces of acculturation and socio-economic change) that are less commonly noted in relation to the state of biodiversity. These socio-economic and cultural factors cause diversity loss by transforming people's relationships to their natural environment. Changing livelihoods, worldviews and value systems alter people's sense of place and cultural identity and lead to a breakdown in the intergenerational transmission of local knowledge, practices and languages that are so closely tied to the surrounding environment. In turn, this has a negative impact not only on cultures and cultural diversity, but also on biodiversity. Of particular interest, because it has rarely been discussed in the context of biodiversity loss, is what causes the loss of local languages. Some of the reasons that sourcebook contributors gave for language loss include: the replacement of indigenous languages by a dominant language, the passing of the older generations who are fluent in the ancestral language (which means fewer opportunities for younger generations to learn the language), intermarriage with immigrants, the actual neglect of indigenous languages in spite of the presence of official bilingual and intercultural education programmes, and the effects of colonialism.

This analysis importantly underscores that biodiversity and cultural diversity are interrelated not only in terms of the factors that account for their synergies, but also in terms of those that lead to the demise of both diversities. The drivers of change may be either direct or indirect, but their impact is invariably the same: the erosion of diversity.

Box 5.1 Industrial Agriculture and Loss of Biocultural Diversity

In Costa Rica (18), the industrialization of agriculture, with its demand for high-yield, homogeneous products, is driving the loss of traditional crops along with the loss of traditional agricultural knowledge and practices. This kind of agriculture is supported by state policies that promote the use of monocrops. These new practices have been a disincentive for the continued cultivation of lower-yield (although nutritionally superior) traditional crops for the maintenance of traditional practices associated with traditional crops. However, a new sensitivity toward biodiversity, a new interest in organic farming and the increased cost of chemical fertilizers and pesticides have all been incentives for a renewed interest in recovering traditional seeds, exchanging them and sharing the associated knowledge.

In Peru, the 'Andean Project for Peasant Technologies' (15) is working to counter negative effects of industrial agriculture and governmental agrarian policies. The local NGO PRATEC supports the recovery of local approaches to agriculture.

Also in the Peruvian Andes, as reported in 'Worlds of Difference' (30), there is pressure for local farmers to join the market economy and replace potato and other crop varieties with more marketable (or higher-yield) 'improved' varieties, or to concentrate on growing a smaller number of traditional varieties for sale to niche markets, instead of continuing to grow a larger diversity of traditional varieties for local consumption. This pressure is threatening agrobiodiversity along with the associated local knowledge.

The contributors to this sourcebook acknowledge that challenges to both biodiversity and cultural diversity are global challenges that are felt everywhere at the local level, and point out that there is an urgent need to address the forces of globalization (changing economies and value systems that induce ecological and socio-cultural changes) on both local and global scales. This is something that most of the projects reviewed here have in common: they seek to address the loss of diversity at a local level, and these efforts ultimately affect biocultural diversity on a global level.

Boxes 5.1 and 5.2 highlight projects that illustrate, particularly well, the specific forces of change that are having an impact on diversity. Box 5.1 showcases projects that are working with the challenges of a loss of agricultural biodiversity, and Box 5.2 highlights the impact that social change in indigenous communities has on an endangered species in the Philippines.

Box 5.2 Acculturation Processes in the Philippines

Massive social and economic change, along with drastic environmental transformation, has engendered profound change in the traditional beliefs, knowledge and practices of Agta and Kalinga peoples in the Northern Sierra Madre on the island of Luzon, Philippines (9). Local people had traditionally been knowledgeable about the behaviour and ecology of the local crocodile species (*Crocodylus mindorensis*) and its wetland habitat, and had passed down this knowledge – as well as the associated beliefs and practices – to the younger generations through stories, myths, taboos and traditional ceremonies. Changes in the livelihoods, education and culture of local people through 'modernization' and acculturation into mainstream Filipino society, as well as massive immigration into the region, have contributed to eroding traditional forms of knowledge about biodiversity in general and about the crocodile in particular, by exposing local people to different belief systems and practices that often appeared to belie older beliefs. The critical reduction in crocodile populations, due both to over-hunting for commercial purposes and to massive degradation and conversion of the crocodile's habitat, has in turn contributed to making the relevant knowledge, beliefs and practices obsolete, by reducing people's familiarity with crocodiles.

Linkages made in biocultural diversity projects

In analysing the projects in terms of the connections they make between biodiversity, cultural diversity and linguistic diversity, we have grouped them into three clusters, according to their main 'entry point' for integrated conservation. These entry points are:

- Biological diversity: The *conservation of biological diversity* achieved by supporting or reviving local cultures and languages or elements of these that ensure (or ensured in the past) biodiversity conservation.
- Knowledge, practices and beliefs: The *maintenance or revitalization of cultural knowledge, practices (management and use) and beliefs* associated with the conservation of biodiversity.
- Languages: The *maintenance or revitalization of local languages, or aspects of a language* that embody information about the natural environment.

Table A1.1 in Appendix 1 provides a summary of the 45 projects according to these entry points and the linkages made among these aspects in conservation objectives. The majority of projects fall into the second two categories, insofar as they emphasize the importance of traditional knowledge, practices, beliefs and languages for biodiversity conservation. They are mainly involved in reaffirming people's connection to the biophysical environment, so that the sense of 'place' and place-based identity is re-established or strengthened. Projects that fall into the first category emphasize biodiversity conservation, but at the same time the approach they take reveals that conservation objectives are difficult to achieve without taking into account cultural beliefs, knowledge, practices and languages associated with biodiversity. Sometimes the distinction between the entry points is blurred, and assignment of a project to one or other entry point is somewhat arbitrary. In some cases, the projects overlap all three categories, in that they systemically integrate and attribute equal weight to cultural affirmation, language revival and biodiversity conservation.

There are also a few projects that do not fit well into any of the three categories, although they still have biocultural conservation as the intended outcome. These include in particular projects that mainly focus on understanding the cultural specificity of worldviews. For instance, the project 'Biocultural Diversity: Elaborating Theoretical Issues for Communities and Policy Makers' in Australia (5) recognizes the need to understand local and indigenous ecological knowledge in order to understand biodiversity, but addresses the issue by looking at how the use of computer databases developed by aboriginal users themselves might assist in developing and enhancing the collective memory in indigenous communities. Another example is the Environmental Applications Reference Thesaurus (11), which examines the cultural dimension of environmental knowledge and how knowledge is organized – something that is important when different cultures with distinct ways of looking at the world work together to address global environmental problems. Projects such as these take 'meta' approaches that are one step removed from making biocultural linkages on the ground, although the linkages are central to the projects' conception.

Overall, the projects in this sourcebook illustrate a variety of ways in which efforts to conserve biodiversity benefit from efforts to support and affirm aspects of local cultures and languages; and conversely, ways in which efforts at cultural affirmation benefit from being linked with biodiversity conservation. From a biocultural perspective, no matter what the 'entry point', these efforts are one and the same, given the interconnectedness of nature and culture. This interconnectedness is expressed in some of the indigenous views articulated in the projects. For example, Rhonda Brim, Aboriginal Native Title Holder in the Wet Tropics World Heritage Area in North Queensland, Australia, makes it clear that the very idea of separation between nature and culture is an artefact: '[There is] no difference, they both together, nature and culture... That's whiteman identifying and dividing nature and culture. When we look at the World Heritage Area we don't just see trees, we see bush tucker, we don't just see rainforest, we see our home, our traditional country' (Pannell, 2006, p72).

Conserving biological diversity through cultural affirmation

Sourcebook projects that have the conservation of biodiversity (including agrobiodiversity) as their entry point also see the need to incorporate and strengthen those elements of culture that are closely tied to local biodiversity. The approach of the 'Bamenda Highlands Forest Project' in Cameroon (4), for example, was not to impose much-needed forest conservation on the local communities, but instead to facilitate a process of consensus building on forest use and conservation based on traditional forest uses and management. Communities were advised on and assisted with practical and legal measures to protect their forests and to help resolve conflicts, while at the same time they were provided with sustainable economic alternatives. The 'Conservation in

Managed Indigenous Areas' project in Ecuador (8) also had biodiversity conservation as its primary goal, but in order to achieve this, it sought to ensure legal rights for indigenous peoples over their ancestral territories and strengthen indigenous identity. The premise was that doing so would afford a greater incentive for the local indigenous communities to conserve biodiversity. The conservation strategy for the endangered Philippine crocodile (9) relies on reviving the traditional beliefs and practices related to crocodiles, which had previously worked to preserve this species. In the central Andes of Peru (15), efforts to stop the genetic erosion in the diversity of native cultivated plants and their wild relatives draw upon and foster a traditional worldview that, in the past, had always led to respect and affection for the natural world. The seed exchange project in Costa Rica (18) addressed the loss of agrobiodiversity – due to government policies that encouraged monocultures – and helped recover the traditional practice of exchanging diverse local varieties of seed among small farmers. In the Xingu Indigenous Park in Brazil (24), which is part of a nationally protected area, conservationists have been involved with Xingu communities in designing a culturally appropriate 'life plan' to conserve the indigenous territories and protect the populations living within them. The endangered wolverine study in northern Canada (26) drew upon aboriginal traditional knowledge of this carnivore species, on which there is limited scientific knowledge, in order to contribute to assessing the status of the species in the context of the national Species at Risk Act. In the 'Local Level Ecosystem Assessment in India' project (33), documentation of traditional knowledge contributes to the People's Biodiversity Register

Box 5.3 Countering Marine Biodiversity Loss Through Reliance on Traditional Cultural Practices

On the island of Lihir in Papua New Guinea (27), local fish stocks are dwindling. Low-impact resource extraction is encouraged in the attempt to reduce over-exploitation of these fish stocks. Project researchers approach the problem of resource depletion by working with local communities to understand and reinstate traditional fishing techniques, tenure systems and use and management strategies within the context of the belief systems that guided them, such as customary restrictions on marine species exploitation associated with local ceremonies and taboos, all of which helped to maintain healthy fish populations in the past.

In the Solomon Islands (12), the Roviana and Vona Vona Lagoons and adjoining coastal zones encompass a number of critical, biodiversity-rich habitats and species, which require protection. Permanent Marine Protected Areas are being established throughout the region through the development of a conservation plan in collaboration with the local communities, which combines customary management and modern conservation methods. Protecting coral reefs by placing the reef systems off-limits to local communities would not be possible without working closely with these communities. The establishment of Marine Protected Areas based on traditional practices and traditional authority structures is the approach that has proven to be the most effective.

database, which has been prescribed by national legislation on biological diversity conservation. In the Southern Rift Valley of Ethiopia, an area known as one of the hotspots of biodiversity, the Konso and Hamar peoples (45) are actively involved in the documentation of indigenous knowledge associated with the use and conservation of biodiversity associated with home gardens, traditional agriculture and sacred forests. Box 5.3 highlights two projects that rely on local knowledge and practices to accomplish the main goal of conserving marine biodiversity in the South Pacific.

Reviving and supporting cultural knowledge, practices and beliefs associated with biodiversity

In our sample of biocultural diversity conservation projects, the projects that highlight the importance of reviving and supporting cultural knowledge, practices and beliefs tend to be the most integrative in their approach to conservation, insofar as they place significant emphasis on cultural affirmation in the pursuit of overall biocultural diversity conservation. In these projects, two key goals can be identified:

1. reviving specific knowledge and practices, including secure land and resource tenure;
2. supporting traditional belief systems as a basis for biodiversity conservation.

Of the projects that emphasize the revitalization of traditional cultural practices and local ecological knowledge, nearly half focus on bringing traditional knowledge and traditional resource management practices to bear on new strategies for land use and biodiversity conservation. For example, the 'Gwich'in Place Names and Traditional Land Use' project in the Northwest Territories, Canada (14) incorporates Gwich'in place names and stories associated with trails, traditional campsites, graves, historic sites, harvesting locales and sacred or legendary places into the Gwich'in Land Use Plan, which is implemented by the territorial government. This ensures that Gwich'in land use is sustainable and respects the local culture. The Tado people of East Nusa Tenggara, Indonesia (23) have documented their knowledge of over 600 ethnobotanical practices involving 200 plant species, in order to guide the conservation of these species. In communities of the Eastern Cape, South Africa (25), it is the cultural value of wild resources, in addition to their utilitarian value, that is driving conservation through continued use in meaningful cultural practices, even in urban contexts. The 'Whitefeather Forest Initiative' in Ontario, Canada (29) combines customary indigenous resource stewardship practices and management tools, rooted in a rich indigenous knowledge tradition, with new forest-based livelihood opportunities for the Pikangikum First Nation youth. In Arizona, US (39), indigenous Apache students are working with elders to revitalize the knowledge of culturally and ecologically important sites, including major springs and wetlands that provide water for the community and sustain a diverse biota. This knowledge will help guide community decisions about ecological restoration of their lands for present and future generations. Similarly, the project in the Sierra Tarahumara, Mexico (6) emphasizes community control over the ecological and cultural health of their landscape and communities. This goal will be achieved through capacity building for ecological restoration and community resilience and through development of educational materials on eco-cultural health that will combine traditional and scientific knowledge. In the project 'Mapping Aboriginal Cultural Values in the Wet

Box 5.4 Securing Resource Tenure for Conservation in Latin America

Four projects in Latin America – two in Brazil, one in Ecuador and one in Costa Rica – work with indigenous communities in their traditional territories. These projects emphasize the need to secure indigenous tenure within these territories as a starting point for ecosystem management and biodiversity conservation.

The project in the Xingu Indigenous Park (24), in Brazil, worked over an 11-year period to abate the frequency of illegal incursions into the park and establish a culturally appropriate management scheme for the park's 4000 indigenous inhabitants. Mapping of the traditional territories, sacred sites, fishing and hunting locales, and other salient features of the landscape has helped drive the conservation of biodiversity in the park.

Also in Brazil, the indigenous peoples of Acre (28) have been trained to sustainably use and conserve resources on the lands that were once taken away from them in government-supported land privatization programmes.

In Ecuador, the goal of the 'Conservation in Managed Indigenous Areas' project (8) was to conserve biodiversity in indigenous territories by focusing on territorial consolidation and legal rights on indigenous ancestral lands, creating capacity for conservation and achieving financial sustainability.

The 'Support Project for the Ngäbe Indigenous People' in Costa Rica (20) broadly focused on assisting the Ngäbe in reversing the loss of their culture by supporting the defence of their territory and the recovery of appropriate management practices, with a particular emphasis on sustainable production and use of traditional medicines, based on the priorities identified by the Ngäbe. A central aspect was the building of organizational capacity and leadership among the Ngäbe.

Tropics World Heritage Area' (42), the project approach is based on the view that the landscape is a biocultural landscape, with biological diversity intricately tied to a diversity of Aboriginal knowledge, values and practices developed over generations. This implies that the landscape cannot be properly protected, managed and perpetuated without taking into account those generations of Aboriginal knowledge, values and practices.

Several projects emphasize the importance of secure land and resource tenure for conservation. These projects embrace the concept that, when local populations have self-determination and security of tenure and are able to govern their own lands, the likelihood of sustainable resource use decisions increases (Colchester et al, 2004). They also support the idea that, when land and resource governance systems are customary systems, it is more likely that conservation will occur (Esmail, 1997; Tucker, 2004). Many of the sourcebook projects are working toward this end. For example, 'Forests and Oceans for the Future' in British Columbia, Canada (13) is a research project that documents ecological knowledge and facilitates the use of customary forms of governance among the Gitxaała First Nation, which will lead to long-term social and ecological sustainability within their traditional territories. The 'Ethnocartography and

Self-Demarcation of Indigenous Peoples' Lands in Venezuela' (35) is a community-based mapping, documentation, training and action project to produce the cartographic, demographic and cultural-historical documents needed to support the land claims of the Hotï and Eñepa peoples. The project is working towards securing legal ownership and title to the lands occupied by these two groups, who are currently faced with pressures for social, technological, economic and ideological change. This goal contributes to the maintenance of both biological and cultural diversity, as well as of the crucial relationships between them, by seeking to secure exclusive rights to land occupation and use for these groups. Box 5.4 highlights four other projects in Latin America that focus on tenure security for biodiversity conservation.

A number of projects focus on another aspect of culture: belief or value systems that are tied to conservation. Several projects work to counter or prevent the loss of cultural beliefs and practices associated with traditional crops, system of taboos related to sacred natural areas, and the use of medicinal plants. For example, ensuring the continued practice of traditional festivals and life-cycle ceremonies related to local crop varieties in the hills of Nepal (17) helps maintain agrobiodiversity. As long as the value system and traditional practices continue to be a part of the social system, traditional landraces are likely to be maintained on farm – an illustration of how agrobiodiversity conservation is possible in culturally rich agroecosystems. In East Nusa Tenggara, Indonesia (23), local knowledge of biodiversity held by the Tado and Waerebo communities is being lost due to, among other factors, conversion to Catholicism. This has resulted in the loss of nature-based rituals related to sacred trees and stone monoliths and of knowledge about plant and animal products used in rituals. The rising price of sacrificial animals, such as pigs and oxen, also makes it more difficult to perform traditional rituals. Knowledge of the more formalized, ceremonial (*adat*) language is in decline. However, documentation of plant and animal species and of the related traditional knowledge and practices in the local threatened Kempo Manggarai language and sharing of this information across the community are helping revive nature-related beliefs and the traditional stories, songs, narratives, customs and ceremonies. Among the Nisga'a nation in Canada (38), ancestral teachings that create spiritual balance are being revived, as they are considered essential in maintaining a reciprocal relationship with the natural environment. The indigenous Gamo communities in southern Ethiopia (43) have a long history of veneration of sacred sites (sacred natural forests, burial grounds, ponds, streams), which remains an effective means to conserve local areas of high biodiversity. The 'Bamenda Highlands Forest Project' in Cameroon (4) documented key medicinal plants from the forest, whose loss would have had a major impact on the local practice of traditional medicine. The 'Promotion of Traditional Medicine and Indigenous Cultural Research and African Spirituality' project in Uganda (19) is entirely focused on protecting and nurturing the medicinal plants that are important to local traditional healers according to traditional spiritual concepts, beliefs and practices, thus ensuring conservation and the sustainable use of biodiversity by local people, specifically healers who use these plants. The traditional belief systems involved in supporting biodiversity conservation can be very complex, as shown by the example in Box 5.5, drawn from the bottle gourd landraces project in Kenya (37). In this case, the belief system provided the foundation for conserving a high diversity of bottle gourd landraces, although this is now increasingly disregarded by the younger generation – one of the challenges the project is facing in reviving this landrace diversity.

Box 5.5 The Role of Kitete Landraces in Local Ceremonies

In the Kitui District of Kenya (37), the diversity of landraces of the bottle gourd (*Lagenaria siceraria*), locally called *kitete*, is maintained by the vast symbolic and cultural value as well as the diverse traditional uses of this species. One *kitete* story says that the ancestral spirits were always the first to plant and the first to eat. Among the Kamba people, the belief is that all dead ancestors live as spirits in a place called '*Ithembo*', which was either a big tree or a rock cave. These ancestors eat like ordinary people; they also have emotions and when hungry or angry they can bring calamities to the community such as rain failure and diseases. In Kitui district, the new harvest is normally accompanied by disease epidemics such as malaria, so sacrifices are made just before the rainy season. Farmers collect all crop seeds and their varieties together – cowpeas, maize, pearl millet, an edible type of *kitete* (*mongu*), a container type of *kitete*, pumpkins, finger millet, sorghum, etc. Elderly women take these seeds obtained from different farmers to the *Ithembo* and offer them as a sacrifice to the ancestral spirits in order to appease them and ensure a rainy season with a good harvest. The sacrifice is also thought to bring blessings to the planting activity. A traditional healer (*mundu mue*) leads the women, advising them on what to do. At the place of sacrifice, the women form a circle and then pour a mixture of all the seeds in a shallow hole (or in a ceramic pot) while uttering a prayer: 'We have brought seeds to you ancestors so that those other seeds we are planting be good seeds. If they will be good seeds we will sacrifice for you in the next season. But if they will not be good seeds, we will not sacrifice to you again.' At the end of the ceremony, they all burst into song as they walk back home. After the crop grows to a stage when it can be consumed, the same elderly women take samples of all these foods to the *Ithembo* for sacrifice. It is believed that by doing this, the new harvest would be blessed and no bad incidences such as diseases would afflict the community. The idea is to make sure that the spirits (who were first to plant) would also be first to eat and therefore there would be no conflicts between the farmers and spirits (story told by Mrs Katheke Mwangangi and MrsWayua Kyalo of the Kyanika Adult Women's Group (KAWG), a farmer's group in Kyanika village, Kitui District, Kenya).

Sustaining and revitalizing languages and associated knowledge of biodiversity

Several of the projects in this sourcebook emphasize the maintenance or revitalization of languages either as their main goal or as an important component of the project. These projects point to the fact that languages hold culturally specific information related to knowledge of local biodiversity and that, when the maintenance of local languages is jeopardized by rapid social and economic change, so too is the local environmental knowledge that the languages encode. The loss of knowledge then puts at risk the conservation of local biodiversity. Conversely, loss of biodiversity contributes to making local environmental knowledge irrelevant, which then contributes to the loss of significant aspects of the language that encode knowledge of biodiversity. When languages are threatened, so is biodiversity, and the cycle continues. The projects exemplify some of the innovative approaches taken to counter language and associated knowledge loss.

The 'Jaru Ethnobiological Language Knowledge Project' (16) emphasizes the role of an Aboriginal-run language centre in supporting an intergenerational transmission of language and ethnobiological knowledge. Elders work with youth to pass on ethnobiological knowledge encoded in the Jaru language. In this way, the project aims to provide language immersion at the same time as it ensures knowledge transmission through a connection with 'place', thus also fostering respect for the land and a stewardship ethic. In the Sierra Tarahumara, Mexico, formal education in schools de-emphasizes the Rarámuri language and ignores Rarámuri culture. The project 'Eco-cultural Health in the Sierra Tarahumara' (6) aims to integrate Rarámuri language, culture and traditional knowledge in alternative in-situ education initiatives within an eco-cultural health framework. The 'Support Project for the Ngäbe Indigenous People' in Costa Rica (20) strived for the revival of the Ngäbere language by producing books and CDs on traditional medicinal plants, oral history and traditional songs, and training Ngäbe youth to write in their mother tongue. This involved coordination between elders and the youth, as well as with the Ministry of Public Education and teaching staff. The 'Ethnocartography and Self-Demarcation of Indigenous Peoples' Lands in Venezuela' project (35) recognizes that ethnobiological taxonomic knowledge is formally encoded, organized and transmitted through language. The project documents changes in language in order to examine links between language, traditional knowledge and environmental change. In the Kitui District of Kenya (37), loss of specific terminologies associated with gourd landraces signals biodiversity-specific language loss. The project gathered information through interviews and by recording songs and stories in the local Kamba language. *Kitete* landraces are also described in the local language, using approximately 70 different names. The 'Vanishing Voices of the Great Andamanese' project in India (40) has gathered audio-video recordings, oral texts and sociolinguistic sketches for the 50 remaining Great Andamanese people. The lexicon pertaining to ecological knowledge of flora and fauna, names and uses of medicinal plants, as well as terms related to hunting and gathering, forms a major part of a trilingual (Great Andamanese–English–Hindi) dictionary. In Hawaii (41), a consortium of educational partners provides a full range of Hawaiian language and culture programmes, from language immersion pre-school to graduate courses at the College of Hawaiian Language (University of Hawaii, Hilo). Native Hawaiians can be educated completely (from 3 months of age up to doctoral level) through a native paradigm, which makes strong connections with the environment. Research in the West Usambara Mountains in Tanzania (44) explores the links between indigenous languages (Kimbugu and Kisambaa) and plant use by studying the indigenous classification system, knowledge and practices around certain local plants, how this knowledge is transferred across generations, and the effects of language shift on traditional environmental knowledge and biodiversity conservation.

The 'Whitefeather Forest Initiative' (29) seeks to maintain the vitality and strength of the indigenous language, as well as the culture and knowledge tradition of the Pikangikum First Nation, within a new economic and resource management context. At the global level, the 'Environmental Applications Reference Thesaurus' project (11) is compiling a multicultural thesaurus with themes related to environment, ecology and biodiversity conservation. The project makes the critical link between the conservation

Figure 5.2 *A Tanzanian elder helping with a plant use consensus analysis*

Credit: Samantha Ross

of languages and conservation of biodiversity, based on the premise that each culture has its own worldview and vision of reality, which is reflected in the language used to talk about the local environment. The 'Medicinal Plants of Antiquity' project (34), which records therapeutic plant use in the Mediterranean region from Classical antiquity to the Middle Ages, also contributes to making this link, through the documentation of ancient languages and knowledge. In doing so, it seeks to better understand human–environment adaptations in the past in order to provide insights on how these adaptations may be useful in the present and future.

These and other sourcebook projects point to the crucial importance of 'language as the missing ingredient of biodiversity conservation' (to borrow the title of project 44), an aspect that until recently had been given very limited attention even in many of the more integrative conservation efforts (Maffi, 1998). In this connection, the rice landraces project in Nepal (17) contributes a word of caution, reminding us that language revitalization by itself cannot ensure the maintenance and transmission of traditional knowledge and practices, which are also critically dependent on the persistence of the relevant belief systems within the socio-cultural context. However, as the projects mentioned above show, ethnobiological knowledge is encoded, organized and transmitted through language, and therefore loss of ethnobiological terminologies signals a loss of biodiversity-specific concepts. Likewise, language loss contributes to loss of traditional stories, songs and rituals relevant to biodiversity. All this strongly indicates that language maintenance or revitalization is an integral component of efforts to achieve biodiversity conservation through cultural affirmation, as well as to bolster local cultures and cultural identity through the protection of local biodiversity.

Collaborative efforts at biocultural diversity conservation

Most of the 45 projects in this sourcebook are highly participatory and collaborative in nature, and several were initiated and/or are led by the communities that directly benefit from them. The information on level of community participation for each project is summarized in Table A1.2, Appendix 1. Some examples are highlighted here to illustrate cases of community-initiated and community-driven projects, followed by examples of projects that, while initiated and conducted from the outside, were significantly participatory.

The Rarámuri people in the Sierra Tarahumara, Mexico (6) saw the need to take action in their communities and recognized the benefits to working with outsiders who could provide needed expertise and other resources. The project was developed as a partnership between two Rarámuri communities and the international NGO Terralingua, based on a mutual relationship that had been building for several years, and on following traditional decision-making protocols. This involved customary consultations between traditional Rarámuri authorities and their respective communities before an invitation to collaborate was issued.

The 'Jaru Ethnobiological Language Knowledge Project' in Western Australia (16) was initiated by the Kimberley Language Resource Centre, an Aboriginal-run body that supports Aboriginal languages in the Kimberley region. The project works with Jaru speakers to identify what kind of resources they believe will best help them pass on their ethnobiological knowledge in their language. The 'Tado Cultural Ecology Conservation Program' in Indonesia (23) has been run by the Tado and Waerebo communities for several years, with financial, administrative, logistical and technical support from the international NGO ECO-SEA. In Brazil's Xingu indigenous reserve area (24), 14 tribal groups formed a partnership with national ministries and the Amazon Conservation Team to develop a culturally appropriate management plan for the Xingu Indigenous Park. The 'Wik, Wik-Way and Kugu Ethnobiology Project' (31) in Australia is a cross-cultural, collaborative initiative between Western-trained scientists and local experts who belong to the Wik, Wik-Way and Kugu Aboriginal groups. The project 'Mapping Aboriginal Cultural Values in the Wet Tropics World Heritage Area' (42) emphasizes that it is critical for Traditional Owners in the region to participate from the beginning in the design and direction of appropriate cultural heritage information management systems in the heritage area. Therefore, the project designed a workshop to empower Traditional Owners so that they can advise on the development and implementation

Figure 5.3 *Extensive community consultations and an explicit mutual agreement were the preliminaries for collaboration between the Rarámuri of the Sierra Tarahumara, Mexico, and an international NGO*

Credit: David J. Rapport

of such systems. The study with Gamo elders on sacred sites in southwestern Ethiopia (43) is based on a cultural movement that is emerging at the grassroots level and among academic institutions and an NGO whose focus is to recapture 'whole indigenous landscapes' and their belief systems, where project success is considered to depend on the participation of ritual leaders, youth and the community at large. Box 5.6 highlights the Whitefeather Forest Initiative, a project completely initiated and led by the local communitiy.

Participatory and collaborative projects that were initiated from outside the local communities include the 'Conservation in Managed Indigenous Areas' project in Ecuador (8), which was undertaken by the United States Agency for International Development (USAID) and designed by a USAID-financed team in consultation with representatives of indigenous federations. Indigenous peoples were fully integrated into the development and implementation of work plans for biodiversity conservation, thus enhancing conservation capacity within indigenous federations. While being a global initiative spearheaded by an international network of conservation professionals, the 'Dance for the Earth and for Her Peoples' project (10) has been developed in a participatory way through direct contact with community groups in different parts of

Box 5.6 Collaborative Partnerships in the Whitefeather Forest Initiative

The 'Whitefeather Forest Initiative' (29) is led by the Pikangikum people of northern Ontario and Manitoba, Canada, and guided by elders based on the tradition of 'keeping the land'. In the effort to maintain this tradition and sustain their own interests while harmonizing them with those of others, Pikangikum establish partnerships for education, resource management and business development with outside parties, based on a spirit of cooperation and mutual respect. This approach strengthens desired outcomes, such as maintaining forest cover and biodiversity as well as the vitality and strength of the indigenous language, culture and knowledge tradition of the community, while at the same time creating new forest-based livelihood opportunities for Pikangikum youth.

the world, and a number of local communities have now taken it up enthusiastically. The project 'Ethnobotany of Indigenous People of the Southern Rift Valley and Southwestern Ethiopia' (45), while initiated by staff at the Department of Biology at Addis Ababa University, seeks to identify best practices for working in partnership with indigenous peoples, in order to ensure mutual trust and equal participation for the fair and equitable sharing of benefits accrued. Similarly, some of the projects that were carried out as doctoral theses (such as projects 17, 25, 32 and 44) were primarily designed by academic researchers, but aimed to be collaborative in nature, with the full participation of the communities involved.

Overall, examining the level of local participation in the projects surveyed here suggests that the degree to which local people were involved in conceptualizing and implementing a project positively correlates with other key characteristics of the projects – for example, the extent to which indigenous or local knowledge is recognized as relevant and incorporated in the project, or the amount of emphasis that is placed on the revitalization of indigenous and local languages. In addition, particularly in projects initiated by outsiders, some of the evidence indicates that higher levels of community involvement may help increase project sustainability over time (that is, whether project activities 'take root' locally and continue beyond the project's timeframe and its initial infusion of human and monetary resources) – key aspects that the literature on project evaluation considers a measure of a project's 'success' (Toulmin and Chambers, 1990; Maffi, 2007b). Significantly, in a number of cases, thanks to successful capacity building, on-the-ground activities have ultimately been devolved to local communities, or have continued in some form without external support after the project's conclusion. Notable examples include the Bamenda Highlands Forest Project in Cameroon (4), where there are now about 20 forest management institutions operating independently of any outside assistance. As well, the Xingu communities in Brazil (24) now manage their own land and cultural conservation efforts. After over a decade of work with indigenous organizations, the Amazon Conservation Team passed on to them the assets of its regional field office, while remaining available to these groups in an advisory capacity. In

Indonesia (23), while the Tado and Waerebo communities still consult with ECO-SEA on project activities, they independently establish and manage work programmes in their communities. After conclusion of the 'Support Project for the Ngäbe Indigenous People' in Costa Rica (20), some activities are still ongoing and some participants are now local leaders.

Beyond these general observations, it was not our goal in this sourcebook to conduct a systematic critical comparison of the outcomes of top-down vs. bottom-up and participatory approaches, in terms of their 'success' as defined according to pre-determined criteria of our own. Nor did we aim to follow the projects longitudinally to evaluate their sustainability over time. Rather, we chose to embrace the diversity of projects as valuable in and of themselves, and viewed 'success' as strongly dependent on who defines it and on the timeframe considered. The success of a project may be measurable on the ground by participants themselves using any number of criteria. As judged by project participants according to their own criteria, most of the projects we surveyed were regarded as successful, whether they were initiated by local communities or by outside sources with community participation. How these initiatives will play out in the long term, and whether their efforts will be sustained over time and even inspire similar initiatives elsewhere, remains to be seen. We return to the issue of 'success' in the context of the lessons learned discussed in Chapter 6.

Methods and institutions for maintaining and revitalizing intergenerational transmission of local knowledge and languages

The effective transmission of traditional knowledge, practices, beliefs and languages from generation to generation is central to the maintenance of the cultural identity and vitality of indigenous and local communities. From a biocultural perspective, it has significant implications for the continued use of specific aspects of culture that contribute to biodiversity conservation, and thus to the sustainability of both ecosystems and human communities. The difficulties that arise in efforts to restore knowledge transmission processes once they have been disrupted are well illustrated by the case highlighted in Box 5.7.

Given the significance of the intergenerational transmission of local and indigenous knowledge, practices, beliefs and languages, here we consider the extent to which the projects in this sourcebook encourage, facilitate and foster the development of institutions that support its maintenance and revitalization. Most of the projects incorporate approaches for continuing the intergenerational transmission of knowledge, practices, beliefs and languages related to local biodiversity. These approaches vary, ranging from the use of more 'formal' tools and institutions (such as educational materials produced for school curricula), to the adoption of more 'informal', elder-to-youth, methods for knowledge transmission (for example, strengthening cultural practices and management institutions, as well as capacity building for improved contact and communication between elders and youth). The information on methods for ensuring knowledge and language transmission is summarized in synoptic form in Table A1.2, Appendix 1.

Several projects document local cultural features and produce written and visual materials for use in educational curricula, which are meant to ensure that younger

Box 5.7 The Importance of Knowledge Transmission to Youth in Tamil Nadu

In the past, the Irular people of Northern Tamil Nadu, India (32) followed traditional practices that ensured that local resources were used sustainably. However, the transfer of responsibility for conservation to government departments has resulted in a loss of control of resources by the communities. As a result, people no longer see any advantages in trying to conserve as they did in the past. If the traditional knowledge held by the community disappears and is no longer transmitted to younger generations, the interest in conserving biodiversity wanes as well. Today, the government is trying to turn that around by working with local communities to ensure greater community participation, but there are difficulties with re-establishing traditions of caring. Irular youth in particular show no interest in conservation, as they see the traditional practices and beliefs as not 'modern'.

generations have access to the traditional cultural knowledge of the generations that preceded them. These include the project documenting the flora and fauna in the Marshall Islands (1), which describes and illustrates local plants and animals with names in the local language, in order to revive the traditional environmental knowledge that has been lost among the younger generations. The project on social, environmental and economic sustainability in Papua New Guinea (21) involves local schools in research projects to document the cultural loss from mining projects that are taking place in the area. The research results are given back to the schools in the form of posters, booklets and videos. The 'Ethnocartography and Self-Demarcation of Indigenous Peoples' Lands in Venezuela' project (35) aims to produce educational materials on ethnogeographical, ethnoecological and other cultural knowledge for use by the Hotï and Eñepa communities. The bottle gourd project in the Kitui District of Kenya (37) involves a community-based *kitete* resource centre that uses audiotapes, photos and printed reports, songs, stories and traditional knowledge to document uses of the bottle gourd, in order to affirm and teach this cultural knowledge. The '*Ndee bini' bida'ilzaahi*: Pictures of Apache Land' project in the US (39) creates computer databases, posters, slideshows, videos and exhibits for the tribal museum. The eco-cultural health project in the Sierra Tarahumara, Mexico (6) aims to develop alternative, hands-on curriculum materials for Rarámuri children and youth, emphasizing the Rarámuri language and bringing together traditional and scientific knowledge. The various initiatives under the 'Knowledge and Language Revitalization in Hawaii' project (41) also offer programmes for hands-on learning in the traditional pedagogy and languages from pre-school to graduate level. The project in the West Usambara Mountains, Tanzania (44) is producing books in the local languages to make traditional knowledge available in an easily understandable format, while also highlighting the need for integrated intercultural and multilingual conservation practices. The 'Wik, Wik-Way and Kugu Ethnobiology Project' in Australia (31) developed a database of local ethnobiological knowledge that documents aspects of the local environment, local plant taxonomies and traditional land management techniques (such as the use of fire) that were being lost. The database is being used as an educational tool for youth as well as in conservation and land management. All of these projects show ways in which knowledge and language is recaptured and maintained for younger generations, future generations and even other communities.

Some projects develop and strengthen the capacity of community management and conservation institutions, thus providing effective ways to keep knowledge and practices alive through continued use. Examples include the development of locally run forest management institutions in the Cameroon project (4); empowering village members to enforce environmental legislation in the crocodile conservation project in the Philippines (9); and establishing Community-Based Marine Protected Areas in the Solomon Islands (12), which has reinstituted traditional authority over people's marine resources, generating innovative governance institutions. Customary forms of governance that regulate human action within the environment are also central to the 'Forests and Oceans for the Future' project in Canada (13). Existing conservation and management institutions within indigenous territories are used in the development of a 'life plan' for the territorial management of Brazil's Xingu Indigenous Park (24). All of these management institutions embed the knowledge and language in current practices and are a means of using and transmitting traditions and local languages.

Projects that reinforce specific cultural practices in order to ensure knowledge maintenance and transmission include the agrobiodiversity project in Peru (15), which emphasizes the importance of the continued practice of traditional 'ritual agriculture' and traditions of nurturing plants, with a special emphasis on children learning these practices in schools. The traditional crop landraces project in Nepal (17) points to the importance of maintaining traditional festivals and life-cycle rituals in which diverse landraces of local crops are used, as a way to help maintain agricultural biodiversity. Similarly, the seed exchange project in Costa Rica (18) has helped revitalize the traditional practice of exchanging local seed varieties, and the oral transmission of the associated traditional knowledge, among small farmers and their families. In turn, this is helping rekindle appreciation for this local genetic diversity and the related knowledge, which had before been cast aside under the pressure of 'modernization'. Reinforcing the traditional belief system around sacred forests in the Gamo communities in southwestern Ethiopia (43) is also regarded as an important way to protect local forest biodiversity.

Several projects emphasize the importance of interactions between elders and youth for knowledge transmission. These include the Gwich'in land use project in northern Canada (14), which brings elders and youth together on their traditional lands to experience the natural environment, so that learning occurs from these hands-on experiences. The language resource centre that is central to the 'Jaru Ethnobiological Language Knowledge Project' in Australia (16) strongly encourages the involvement of younger generations in bush trips and other activities to increase their immersion in the language.

The 'Pictures of Apache Land' project in Arizona (39) teaches the Apache youth in the community about traditional Apache knowledge of the land and the historical, social and moral interpretations their ancestors had of places in the landscape, in order to instil in the youth a commitment to ecological restoration. In supporting the development of an alternative educational curriculum for Rarámuri children and youth, the 'Eco-cultural Health in the Sierra Tarahumara' project in Mexico (6) also seeks to

Figure 5.4 *Lydia Ozies in the bush with* mardiwa *(edible gum) in the Kimberley, Western Australia*

Credit: Kimberley Language Resource Centre

foster and strengthen interactions with the elders and thus the continued or renewed transmission of the local language and cultural knowledge. The Ngäbe project in Costa Rica (20) encouraged the recovery of oral history from the elders and the teaching of it in indigenous schools. The 'Whitefeather Forest Initiative' in Ontario, Canada (29) also actively fosters elder-to-youth teachings about the traditional knowledge, language and stewardship values of the community, in the context of developing new forest-based livelihood opportunities for the youth.

Implementing or developing biocultural diversity policy

Directly or indirectly, most of the projects in this sourcebook have implications for conservation policy at some level, whether at the local, national or international scale. Table A1.2 in Appendix 1 summarizes some of the approaches that the projects are taking to either develop biocultural conservation policy or implement existing bioculturally relevant conservation policy.

By and large, these projects do not specifically aim to affect international policy, or have the means to do so. Nevertheless, in some cases they are linked to international policy in significant ways. Several projects follow the international Convention on Biological Diversity (CBD) as a guideline for their work. For example, in the 'Bamenda Highlands Forest Project' (4), the government of Cameroon has supported the revitalization of the important cultural values of these forests because of its obligations as a signatory to the CBD. The project 'Ethnobotany of Indigenous People of the Southern Rift Valley and Southwestern Ethiopia' (45) is developing best practices for access and benefit sharing, in line with the principles promoted by the CBD. The Yunnan initiative, on which the project on indigenous knowledge of Yunnan ethnic minorities (22) is based, also endorses the CBD's call for respect of cultural and spiritual values for sustainable development. Other projects make links with the UNESCO World Heritage Convention (WHC). Examples are the 'Whitefeather Forest Initiative' in Canada (29), one of whose goals is to establish a protected area on Pikangikum territory with UNESCO World Heritage status, and the 'Mapping Aboriginal Cultural Values in the Wet Tropics World Heritage Area' project in North Queensland, Australia (42), which is working for the relisting of the Wet Tropics World Heritage Area as a biocultural landscape under the WHC. Another project associated with international environmental policy processes is the 'Dance for the Earth and for Her Peoples' project (10), which is influenced and endorsed by IUCN's Commission on Environmental, Economic and Social Policy (CEESP) and World Commission on Protected Areas (WCPA). The goal of the 'Environmental Applications Reference Thesaurus' project (11) is to develop an advanced information management tool for use in international environmental research and policy – a context in which many diverse cultures with distinct worldviews need to work together to address environmental problems.

Several projects seek to affect conservation policy at the national level by developing policy guidelines for the protection of indigenous reserves. Examples are the 'Ethnocartography and Self-Demarcation of Indigenous Peoples' Lands in Venezuela' (35) and the 'Conservation in Managed Indigenous Areas' project in Ecuador (8). The latter has been highly successful in strengthening indigenous federations and in legalizing ancestral territories according to Ecuador's laws. Similarly, the 'Support

Project for the Ngäbe Indigenous People' in Costa Rica (20) produced a legal study to influence national policy change with regard to indigenous rights to manage natural resources. The project to establish Marine Protected Areas (MPAs) in the Solomon Islands (12) aims to legalize all MPAs at the provincial and national levels, based on a network of protected areas to conserve marine and riparian habitats. In Peru, PRATEC (15) helped promote the incorporation of local knowledge into the school curriculum and the adoption of the local agricultural calendar as matters of national policy. Both the project in Tanzania (44) that examined the biocultural dynamics of language and the endangered Great Andamanese language project (40) highlight the implications of language shift for biodiversity conservation policy and the need for national-level policy to assist in the revitalization of threatened languages and cultures.

Other projects address policy at state, provincial or local levels. Several of these are located in Canada. The 'Forests and Oceans for the Future' project in British Columbia (13) contributes to policy development and evaluation, by using research results within the provincial government's Land Resource Planning Process. The 'Gwich'in Place Names and Traditional Land Use' project (14) provides input into Northern Territories policies and legislation that concern Gwich'in heritage resources, as well as information for the development of the Gwich'in Land Use Plan for sustainable land use. A Gwich'in traditional knowledge policy, called 'Working with Gwich'in Traditional Knowledge in the Gwich'in Settlement Region', is now in use. The 'Whitefeather Forest Initiative' (29) has developed a Community-based Land Use Strategy for the Whitefeather Forest and Adjacent Areas in the context of the Province of Ontario's Northern Boreal Initiative, which seeks to develop forest management approaches that are ecologically suited to the northern boreal forest. Box 5.8 provides an example of conservation legislation in Canada related to species at risk, which mandates the inclusion of a traditional ecological knowledge component.

In Papua New Guinea, the project on traditional knowledge of marine environment and fishing on Lihir Island (27) seeks to influence local-level policy by promoting traditional low-impact marine resource extraction practices to reduce over-exploitation of fish stocks. The 'Local Level Ecosystem Assessment' project in India (33) contributed to the local implementation of India's national biocultural policy, the National Biological Diversity Act of 2002. The Act recognizes the relevance of traditional knowledge of biodiversity and traditional conservation practices such as sacred groves and sacred water bodies, and mandates the recording of such knowledge and practices in a national People's Biodiversity Register. In this case, the linkage of institutions from the local to the national level works to 'scale up' local knowledge to the level of national policy. At the same time, by recording species' names in the local vernacular and linking them to scientific nomenclature in support of CBD-related claims to Intellectual Property Rights and Access and Benefit Sharing concerning biodiversity, the project contributed to the on-the-ground implementation of international policy.

In various ways, many of the projects in this sourcebook seek to cross scales to make local-level issues relevant to policy at sub-national and national levels, and conversely by making national or international policy relevant to issues at sub-national and local levels. By cutting across multiple layers of decision making and attempting to link local realities to sub-national, national and international processes, these projects are representative of how biocultural-diversity-oriented initiatives can have the most significant impact on

Box 5.8 Using Traditional Knowledge for Policy Decisions

The project 'Use of Aboriginal Traditional Knowledge in Species Assessment: A Case Study of Northern Canada Wolverines' (26) was designed to bring aboriginal traditional knowledge (ATK) to bear on the assessment of the status of the endangered wolverine (*Gulo gulo*) species in the context of Canada's Species at Risk Act (SARA). SARA explicitly recognizes the value of ATK in assessing species status, and the Committee on the Status of Endangered Wildlife in Canada (COSEWIC), which is in charge of evaluating the status of species in Canada, is required by law to incorporate ATK along with scientific knowledge; yet, ATK has rarely been used for this purpose. By researching the case of the wolverine, this project helped close the gap between local ecological knowledge and formal science, so that the former can be integrated into federal regulatory processes to protect endangered species. The study demonstrated that ATK contributes invaluable information regarding the status of species in Canada, and provided recommendations as to how ATK can be documented, described and utilized in COSEWIC's species assessment process. The project also showed that focusing on ATK and the significance of species for aboriginal peoples can significantly foster aboriginal involvement in species conservation, which may ultimately improve local-level acceptance of a species' status and associated recovery programmes.

Box 5.9 Guardians of the Forest

We are the guardians of our lands and our forests. We have been planting several kinds of fruit trees, hardwoods, palm trees and other trees useful to our people and our society. To carry out our environmental management, we would like to work with the environmental inspection [agencies]. [We say] no to destruction, no to misery. One of the jobs of the Indigenous Agro-Forestry Agent is to inspect and protect the native lands from invasions, hunters, professional fishermen, lumberjacks and other people interested in exploiting our environmental natural resources. We would like the state to recognize our profession and the government of the forest and other governments to help us to have more courses here at the Training Centre for the Peoples of the Forest. We need help, commitment and comprehension so that our profession as inspecting agents is recognized. We are starting to replant the indigenous territories with abundance and happiness for all living beings.

Source: from a testimony given by Indigenous Agro-Forestry Agents (2000)

policy. Box 5.9 offers a testimony from the Indigenous Agro-Forestry Agents in Acre, Brazil (28) on how local actors see the importance of influencing state policy, and on the importance of government assistance in these efforts.

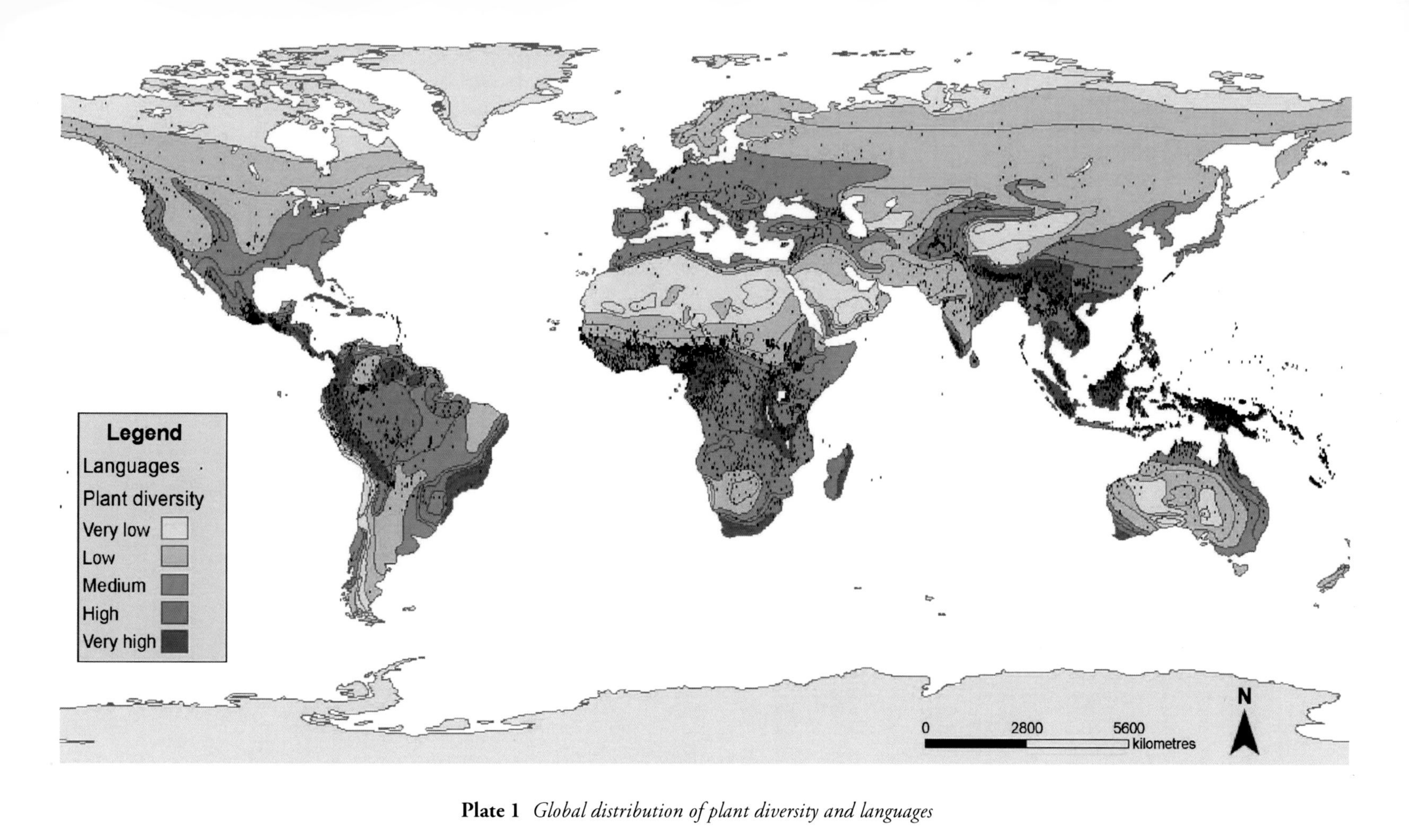

Plate 1 *Global distribution of plant diversity and languages*

Source: From Stepp et al (2004), based in part on data from Barthlott et al (1999). Used by permission of the authors.

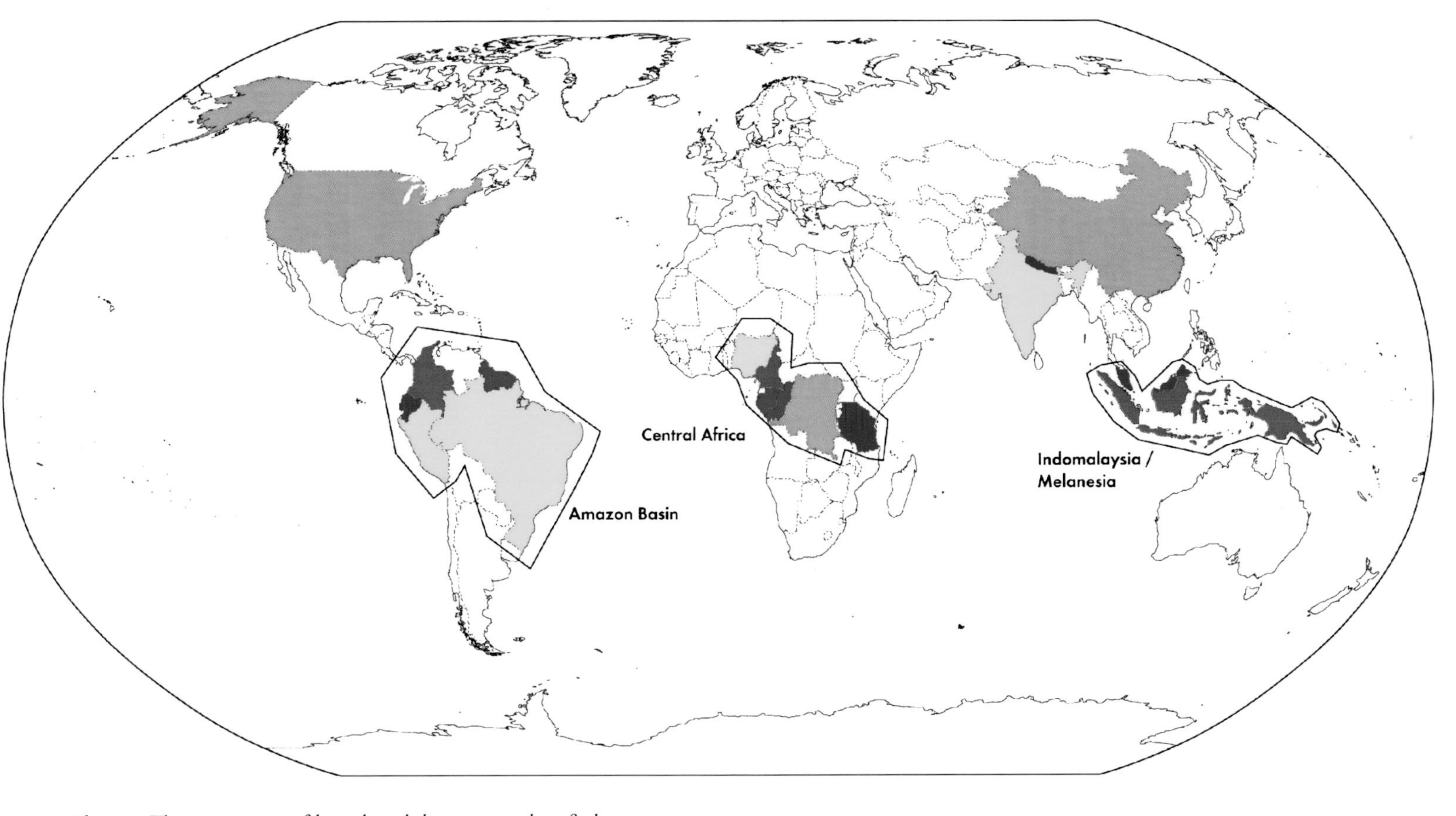

Plate 2 *Three core areas of biocultural diversity as identified by the Index of Biocultural Diversity*

Source: Original map by David Harmon and Jonathan Loh, based on work published in Loh and Harmon (2005). First published in Harmon (2007). Used by permission of the authors.

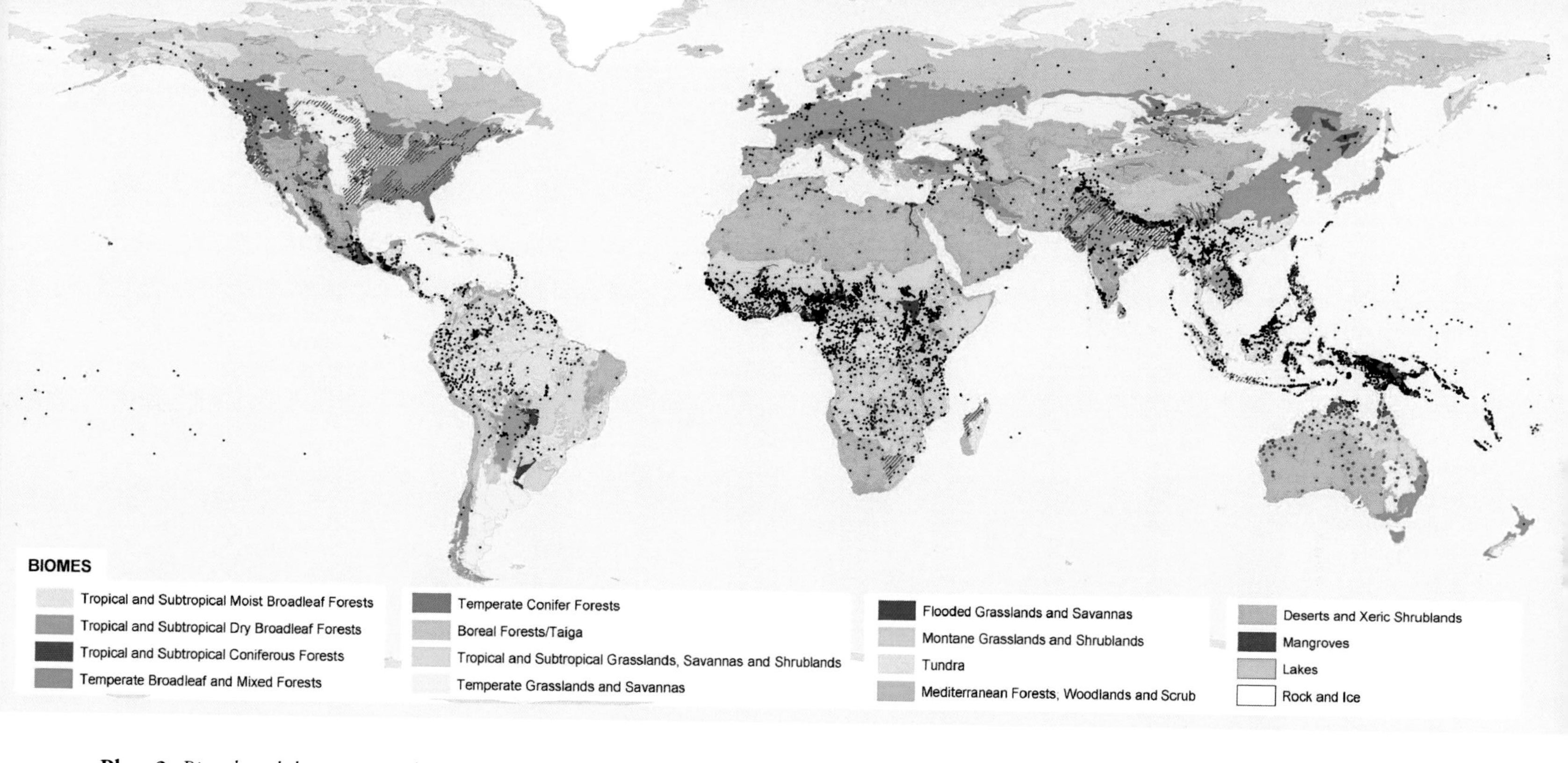

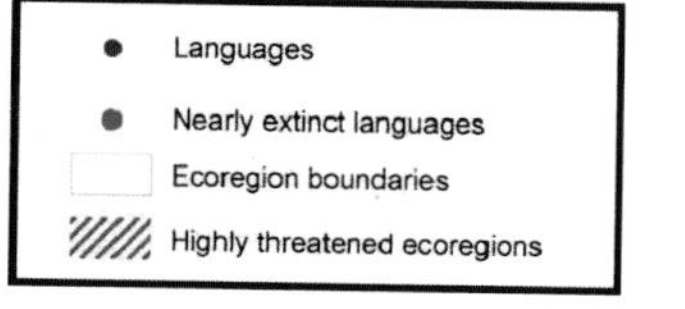

Plate 3 *Biocultural diversity at risk: The world's threatened ecoregions and ethnolinguistic groups*

Source: Languages data from Ethnologue, ecoregions data from WWF, used by permission.
Map prepared by Conservation Technologies Institute for Terralingua (2002).
A poster version of this map was published in Skutnabb-Kangas et al (2003).

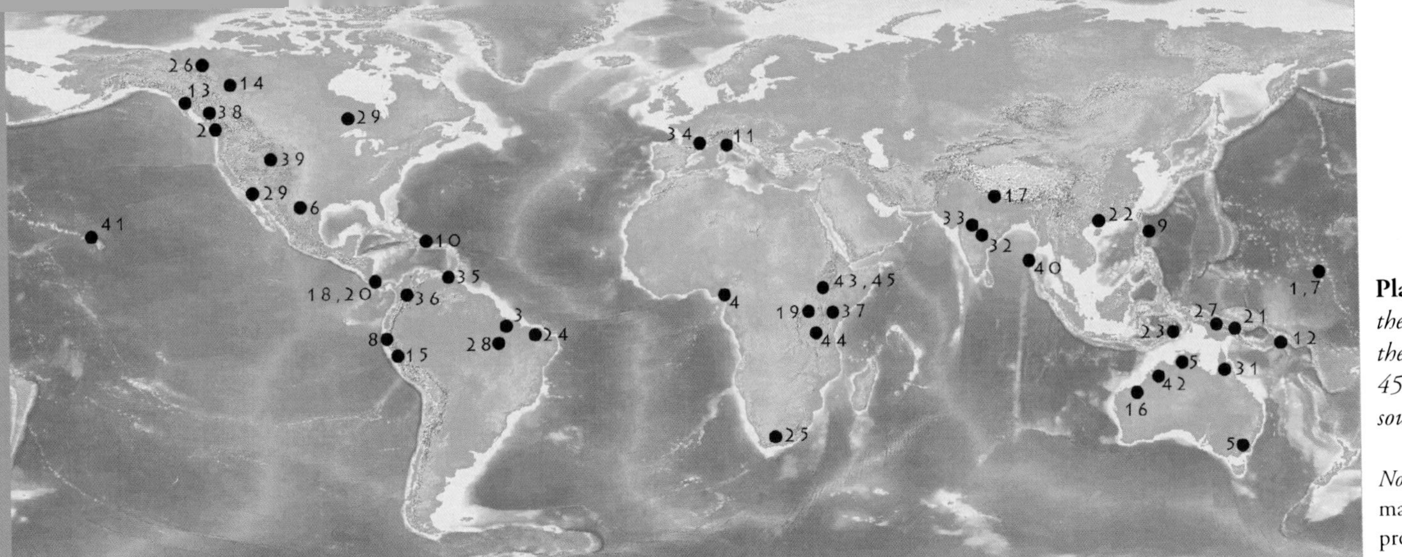

Plate 4 *Map of the world showing the location of the 45 projects in this sourcebook*

Note: Numbers on the map correspond to the project names listed on this plate. The same numbers are used in the main text and in Appendixes 1 and 3 to uniquely identify each project. Project titles on this map have been shortened due to space constraints. For full titles see Appendix 3.

Source: Original work by Ellen Woodley and Ortixia Dilts for Terralingua

Plate 5 *Drawing representing indigenous perspectives of the local environment and visualizing innovative and creative strategies for maintaining the landscape in Yunnan, China*

Source: Xu Jianchu

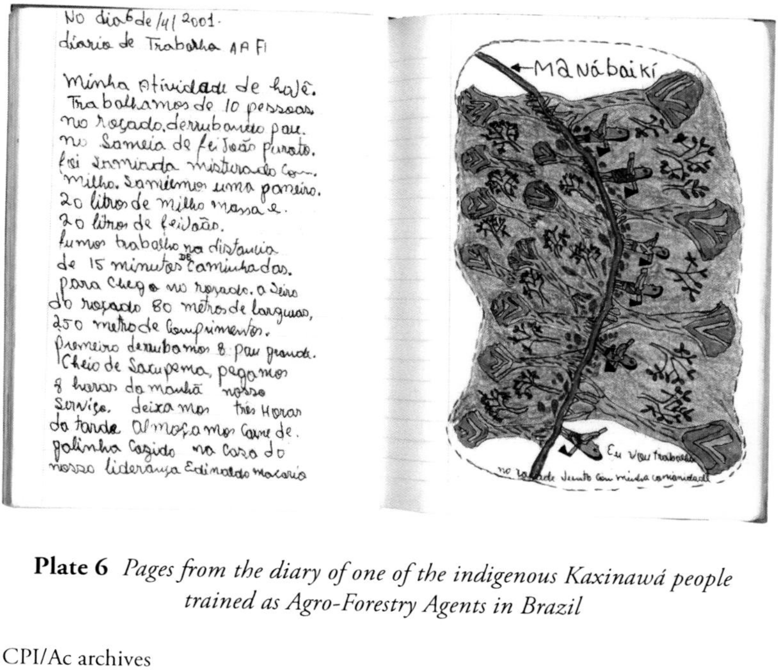

No dia 6 de /4/ 2001.
diario de Trabalho AAFI

Minha Atividade de hoje.
Trabalhamos de 10 pessoas
no roçado, derrubando pau.
no Sameia de feijão purato.
foi Sameiada misturado com
milho. Samiemos uma paneiro.
20 litros de milho massa e.
20 litros de feijão.
fumos trabalho na distancia
de 15 minutos de caminhadas.
para chega no roçado. o Seiro
do roçado 80 metros de larguras,
250 metro de Comprimentos.
Premeiro derrubamos 8 pau grande.
Cheio de Sacupema, pegamos
8 horas da manhã nosso
Serviço. deixamos três Horas
da tarde almoçamos Carne de.
galinha Cozido na Casa do
nosso liderança Edinaldo Macario

Plate 6 *Pages from the diary of one of the indigenous Kaxinawá people trained as Agro-Forestry Agents in Brazil*

Source: CPI/Ac archives

Plate 7 *Boys in their* chacra *(cultivated field) in Matara, Cajamarca, Peru*

Credit: Jorge Ishizawa

Plate 8 *Girls in the highlands in Cusco, Peru*

Credit: Jorge Ishizawa

Plate 9 *Gamo elders praying in Dorbo Meadow in Southern Ethiopia at the beginning of the Mascal ceremony*

Credit: Christopher McLeod

Plate 10 *Documenting the morphological diversity of* kitete *gourds in Kitui District of Kenya*

Credit: Yasuyuki Morimoto/Bioversity International

Plate 11 *Young Agta girl spearfishing in the Disulap River Philippine crocodile sanctuary in the municipality of San Mariano, Isabela Province, Luzon, Philippines*

Credit: Jan van der Ploeg

Plate 12 *Learning by doing is an important method of knowledge transmission among the Eñepa people of Venezuela*

Credit: Stanford Zent

6

Lessons Learned from the Projects

Ellen Woodley

Given the magnitude and implications of the worldwide decline of biocultural diversity, and the multiplicity of factors that are contributing to this decline from the local to the global scale, the projects in this sourcebook collectively represent a remarkable tour de force in working towards solutions to stem global diversity loss. In their varied approaches to addressing this crisis, these projects offer lessons that are relevant for strengthening future efforts at maintaining and enhancing biocultural diversity. In this chapter, we examine some of these 'lessons learned' from the projects. First we focus on the commonalities and the differences in the approaches the projects take to integrated biocultural conservation. We then highlight some of the key conditions that appear to be required for the success of biocultural diversity conservation projects, as well as some of the challenges that these efforts face. In a final reflection on lessons learned, we conclude that local-level projects can make a significant impact, but their long-term success requires continued efforts at the local, regional, national and international levels to address the complexity of factors that contribute to biocultural diversity loss.

Commonalities and differences among the projects

As diverse as the projects surveyed here are, they all have in common the recognition of the critical link between biodiversity conservation and the affirmation of cultural values, beliefs, knowledge, practices and languages, particularly those of indigenous peoples and local communities who live in close association with the natural environment. One sourcebook contributor (Dawn Marsden, project 2) emphatically expressed the nature of the relationship between biodiversity and culture: 'I am convinced more than ever that biodiversity on this planet is inextricable from cultural diversity, and more specifically, from traditional-based cultures.' The contributor further argued that teaching the logic of interconnectedness (a logic that lies at the core of many indigenous worldviews and provides the ethical basis for the field of biocultural diversity) brings along with it concepts of integrity, responsible action and sound relationships, as well as the idea that all of our actions have consequences. In various forms, similar sentiments motivate the work of all the projects reviewed in this volume.

While having that common emphasis, the projects also exemplify a number of different approaches to integrating biodiversity conservation and cultural affirmation objectives in order to strengthen the sustainability of human–environment relationships. As we discussed in the previous chapter, there are three 'entry points' in terms of what each project holds as its main priority: biodiversity conservation; maintenance or revitalization of knowledge, practices and beliefs; and maintenance or revitalization of local languages, or of aspects of a language that embody information about the natural environment. Within these three categories, we also identified some key recurrent approaches:

1 encouraging and strengthening existing traditional knowledge and management practices that contribute to biodiversity conservation;
2 supporting land claims, resource tenure and governance systems to enable locally controlled decision making on sustainable use and management of local biodiversity;
3 building on nature-based belief and value systems and strengthening cultural identity to sustain and enhance local biodiversity;
4 reviving and revitalizing languages or aspects of language that embody knowledge of biodiversity.

Whatever their entry point, the majority of projects incorporate one or more of these approaches. For example, while some of the projects have as their primary goal biodiversity conservation (the first entry point), they recognize the need to take into account relevant aspects of culture as a means to achieving their objectives. All of these projects thus incorporate, to various degrees, one or more of the four approaches mentioned above. Those projects that have as their primary goal the maintenance or revitalization of local or traditional knowledge (the second entry point) are especially likely to also devote attention to supporting traditional management practices, tenure security and belief/value systems, as well as, to some extent, strengthening local languages. Projects that focus on the maintenance or revitalization of local languages, or of aspects of a language that embody information about the natural environment (the third entry point) bring to the fore in particular the intricate links and feedback mechanisms between language, knowledge and the environment. Rapid social and economic change leads to the loss of languages, or of certain elements of languages that embody culturally specific knowledge of multiple aspects of biodiversity (for instance, taxonomic names as part of distinct biological classification systems, names of species habitat, and descriptors of species abundance); in turn, loss of linguistically encoded environmental knowledge has consequences for biodiversity conservation. At the same time, when biodiversity is lost and the related knowledge becomes irrelevant, languages are weakened, which further fuels a cycle of language, knowledge and biodiversity loss. The projects in this category point to the crucial connection between language revitalization and biodiversity conservation: languages cannot be revitalized completely and effectively without paying close attention to how they are tied to local biodiversity through the environmental knowledge they encode; conversely, conserving biodiversity requires fostering the vitality of languages and of the linguistically encoded environmental knowledge. These projects also stress the importance of local languages for strengthening cultural identity, another key component of cultural resilience in the face of rapid change.

Regardless of entry points and approaches, the projects in this sourcebook share another important characteristic: the recognition that integrated biocultural conservation involves a linkage of scales between the local and the global. Project contributors generally recognize that the challenges to sustaining biocultural diversity are at the same time local and global challenges. Forces of globalization bring about cultural and environmental changes at the local level – for example, changes in lifestyles, both imposed and by choice, due to changing economies and value systems, leading to over-exploitation of local natural resources. In turn, these local-level changes form part of the feedback loop that contributes to biological and cultural diversity losses that are ultimately felt at the global level. Conversely, efforts to carry out integrated biocultural conservation at the local level contribute to sustaining cultures and biodiversity locally, which then affects diversity at the national or global level.

Many of the projects are impacted by processes that cross scales from the global or national level to the local level. In addition, some of the projects are able to influence processes that cross scales from the local to the national/global level. An example of crossing scales from the global level to the local level is the Bamenda Highlands Forest Project (4), in which the Cameroon national government, under its obligation as signatory to the Convention on Biological Diversity (CBD), has supported the revitalization of the important cultural values of biodiverse montane forests that otherwise would have probably been lost to extractive forestry practices. Another example is that of the Tado and Waerebo communities in East Nusa Tenggara, Indonesia (23), where the principles embodied in the International Society of Ethnobiology (ISE) Code of Ethics have been applied, so that all ethical requirements for conducting collaborative research with local communities are fulfilled. For the Wet Tropics World Heritage Area project in Australia (42), seeking World Heritage status on an international scale is an incentive to promote the area as a 'biocultural landscape' – a status that will help maintain local knowledge and languages in the Aboriginal communities in the area.

Projects that cross scales by 'scaling up', whereby local activities impact national or global processes, include the People's Biodiversity Register in India (33), in which local knowledge and innovation contribute to a national database; the Marine Protected Areas (MPAs) project in the Solomon Islands (12), where the MPAs are locally managed but contribute to a network of legalized protected areas at the national level; and the endangered wolverine study in Canada (26), which incorporates aboriginal traditional knowledge into a national-level species-at-risk strategy.

Such explicit efforts to make linkages across jurisdictional and geographical scales result in these and other projects having a direct and identifiable impact at multiple scales. However, regardless of whether projects do or do not deliberately attempt to make cross-scale linkages, every local effort to conserve biodiversity and support local cultures is influenced by both local and global forces and thus needs to take multi-scale dimensions into account. At the same time, local efforts to sustain biocultural diversity ultimately have a positive impact on diversity at the global level.

Conditions for success and challenges in biocultural diversity conservation

The main lesson to be learned from the diverse but converging efforts at integrated biocultural conservation at the local scale concerns the conditions that are required for project success. These conditions vary by project, but it is clear that, depending on the particular context of each project, certain requirements need to be in place if biocultural diversity is to be sustained. There are also certain challenges faced by several projects that are context-specific, but lessons can be learned from these as well. Whether it is a breakdown in community cohesion or an impoverished socio-cultural base from which to revitalize the local language, there is knowledge to be gained by understanding the obstacles that people on the ground encounter in their efforts to sustain biocultural diversity.

Based on successes and challenges, as reported in the projects, in this section we discuss some key conditions that should be met and some of the issues that need to be addressed to ensure biocultural diversity conservation. Some of the conditions have more to do with strengthening and enhancing internal institutions and processes, while others are ways to ensure that indigenous and local communities are better able to affect the outcomes of external events that have an impact on their lives. In general, internal and external circumstances in projects tend to interact in multiple ways.

Maintaining and restoring the strength of local institutions

Local institutions give voice to local concerns and empower communities to be involved in decision making on matters that affect their interests, lands and resources, livelihoods, security, well-being and overall way of life. Local institutions that are relevant for biocultural diversity conservation range from traditional governance, socio-economic, cultural and spiritual institutions to newly established ones that are well integrated within a traditional context. Relevant traditional governance institutions are seen at work in the 'Whitefeather Forest Initiative' (29), where community elders take a leading role in guiding the land use planning process, based on knowledge tradition, language and stewardship values of the community. 'Forests and Oceans for the Future' (13) is reviving Gitxaała laws (*Ayaawk*) and history (*Adaawk*) that describe relationships of trust, honour and respect that are appropriate for the well-being and continuance of the people, and that also define the rights of ownership over land, sea and resources within the territory. The Gwich'in Social and Cultural Institute (14) was established in 1992 as a result of concern among the Gwich'in about the loss of their culture and language, and the impact this was having on their families; its purpose is to support and strengthen the traditional institutions that ensure language and culture transmission.

Examples of socio-economic, cultural and spiritual institutions that support biocultural diversity include the 'whole indigenous landscapes' movement that is taking place among the Gamo people in southwestern Ethiopia (43), where the focus is on the conservation potential of traditional belief systems. Similarly, traditional leaders and spiritual beliefs maintain sacred forests in the southern Rift Valley in southwestern Ethiopia (45), while the Nepalese rice culture project (17) revolves around festivals and life-cycle rituals that maintain diverse local varieties of rice.

Among the 'new' institutions that help sustain and improve biocultural diversity are the community-based forest management institutions in Cameroon (4), which have a number of roles relating to the planning and management of the forest and reporting to the government. In the 'Traditional-Based Indigenous Health Services' project in Canada (2), the new inter-Nation council of health practitioners is evolving to fulfil the holistic health needs of some First Nations. Another example is the traditional healers group in Uganda (19), which is comprised of about 100 healers from different areas who meet in a culturally diverse setting at a site that is becoming a traditional health and cultural institute.

Sometimes, newly formed institutions run into difficulties, as in the case of the Indigenous Land Association of the Xingu (24). This association was established as a means for dialogue with national society but has not been considered successful, because of an administrative structure that conflicts with the traditional political structures of Xingu indigenous societies. The institution presumes command of Portuguese, basic mathematics, legislation and inter-institutional relations. It is mostly the youth who have this kind of 'new' knowledge, and this tends to create conflict with traditional village-level institutions that are dominated by elders.

One of the most widespread challenges that the projects encounter in efforts to maintain or revitalize strong local institutions relevant to biodiversity conservation is the breakdown in the maintenance and intergenerational transmission of traditional values, knowledge, practices and languages. For example, the projects in the Sierra Tarahumara, Mexico (6), Yunnan, China (22), the Philippines (9), Peru (15), and several others (e.g. projects 22, 31, 32, 35, 37, 40) all face the consequences of this breakdown. The causes range from exposure to acculturation processes and formal education that diminish the authority of traditional worldviews and teachings, to economic pressures that cause younger people to leave their communities and look for work in cities, industrial farms, or manufacturing plants.

The key issue in this context is how to maintain and strengthen the local (both informal and formal) institutions that support intergenerational transmission of values, beliefs, knowledge, practices and languages associated with biodiversity, so that local communities may be better equipped to cope with the momentous changes that are affecting them. Most of the projects are finding ways to strengthen intergenerational links and reaffirm the value of certain traditions that instil pride in culture, and to impress on people the importance of passing on cultural traditions that will assist in contemporary issues of biocultural diversity conservation. Ways to enhance these formal and informal institutions include the production of educational materials for schools, creating opportunities for elders and youth to work together, and strengthening resource management institutions.

Institutions that ensure the vitality of cultural values, beliefs, knowledge, practices and languages associated with biodiversity may also become relevant at other scales, such as at the level of national policy. Several projects, for example those that seek to strengthen resource management institutions in traditional territories in Venezuela (35) and within the Xingu Federal Reserve in Brazil (24), focus on capacity building to enable traditional institutions to work across scales, from the local to the national and vice versa. In the Peruvian Andes (15), the incorporation of local knowledge into the school curriculum and the adoption of the local agricultural calendar have become

Figure 6.1 *One of the key challenges faced by the Rarámuri of the Sierra Tarahumara, Mexico is the intergenerational transmission of traditional values, knowledge, practices and language*

Credit: David J. Rapport

national policy. In these and other cases, strengthening key institutions can help 'scale up' local traditional knowledge, practices and languages.

Securing land and resource tenure

Several projects, including two in Brazil involving the Xingu indigenous peoples (24) and the indigenous peoples of Acre (28), as well as the projects involving the Ngäbe in Costa Rica (20), the Muisca in Colombia (36) and indigenous peoples in Ecuador (8) and Venezuela (35), all strive to ensure security of land tenure and resource access through land claims based on existing national regulations related to traditional rights to indigenous territories. With the enforcement of indigenous rights, it then becomes feasible to strengthen the capacity of the communities involved and draw upon the local practices that have been shown to sustainably manage local biodiversity.

However, many of the projects share a major challenge in relation to land and resource tenure security: what happens within indigenous territories or customary lands is often profoundly affected by what happens outside of them. In the case of the Xingu Indigenous Park (24), tenure security, and thus the sustainability of the park as a whole, depends on the politics of environment and development outside the park and on the larger picture of defending the biodiversity of the Amazon. This makes it crucial for the Xingu peoples to identify potential allies and to sensitize the relevant public agencies and the public in general to what is happening in the Xingu region. Challenges to the project with the Muisca community in Colombia (36) relate to the perception among non-Muisca that a strengthened Muisca community poses a threat. Non-Muisca feel that the Muisca might challenge the land rights of private landowners in the area, particularly in relation to the Muisca's attempt to recover their sacred sites. Similarly, encroachment by outsiders such as loggers and cattle ranchers threatens ecosystem restoration efforts by the Rarámuri in the Sierra Tarahumara, Mexico (6). Therefore, while these projects emphasize the importance of securing land and resource tenure so that indigenous peoples and local communities can sustainably manage the biodiversity in their territories, they also point to the difficulties that can be encountered in doing so, and to the need to create a more supportive environment for the rights of indigenous peoples and local communities.

Strengthening cultural identity

A strong cultural identity confers resilience to cultural practices, knowledge and languages, which in turn enhance and validate efforts to maintain sustainable livelihoods and protect biodiversity. Strengthening and valuing cultural identity is important in several projects, both for indigenous peoples and local communities and for the policy makers, government agents, conservationists, resource managers and other outsiders who directly or indirectly work with these communities. As one of the sourcebook contributors notes, biodiversity conservation is 'above all a matter of places and localities' (Jan van der Ploeg, project 9). This statement underscores the need for local people to feel a sense of pride in their own locality and the associated biocultural heritage, and for outsiders working with local people to recognize the importance of this sense of pride and to avoid undermining it, whether directly or indirectly.

In the case of the Muisca people in Colombia (36), the strengthening of cultural identity is central to the project: with early colonization resulting in displacement from traditional territories and disruption of traditional culture, the project's imperative is to revitalize and affirm cultural identity by reviving ancestral cosmological knowledge and spirituality, to revitalize the Muisca language, and to restore traditional practices, teachings, and knowledge of the natural world. The Kenyan project to conserve bottle gourd landraces (37) further highlights the importance of recognizing traditional knowledge, which is a major factor underpinning social cohesion, empowerment and human capital in poor rural communities in Kenya that depend on biodiversity for their livelihoods.

Challenges to maintaining or recovering a previously strong cultural identity come from both outside and within indigenous and local communities. The Gamo elders project in Ethiopia (43) found that the values and customs that maintained sacred sites are being disregarded and undervalued by the state, conservation agencies, policies and laws. The main challenges in implementing this project were to work with government officials who lacked respect for indigenous spirituality and the failure of the state to protect indigenous peoples and their religious practices, making it all the more difficult to strengthen and validate the traditional culture. However, with perseverance and awareness raising, the government did, in the end, provide support. In Tanzania (44), where language is a politically sensitive issue, there is a major shift from indigenous languages to English (the former colonial language) and Kiswahili (the language that was promoted as the national language to foster unity, national identity and tribal cohesion after independence in 1961). Local languages are not officially recognized and are banned from schools and the media. The project highlighted one reason to validate and encourage the use of indigenous languages: that these languages are the ones used to convey ethnobotanical knowledge, and that therefore the mother tongue is important for plant conservation and for transferring indigenous plant knowledge across generations. In the case of the language documentation project in the Andaman Islands (40), there was an internal challenge to language survival and cultural identity: most community members held the view – common among many communities under intense 'modernization' pressure – that losing their language is not a problem, as they did not think of their language as having any relevance to the modern world in which they now live. Recapturing the language in a dictionary, the project contributor reports, may rekindle Andamanese pride in their ancient tongue.

Reconnecting elders and youth

As we saw in Chapter 5, in many indigenous societies and local communities, socio-cultural and economic change has disrupted the relationship between generations and has caused a disjuncture in transmission of traditional cultural values, beliefs, knowledge, practices and languages. In many of the projects, the breakdown or weakening of institutions for intergenerational transmission, as well as the changing value systems of youth, point to the importance of strengthening the relationship between elders and youth and of reviving interest and pride in culture, language and 'place' among the youth. A renewed interest in traditional teachings and biodiversity-related knowledge can be an incentive for the younger generations to sustainably use and conserve local biodiversity. The projects employed various means to reconnect elders and youth, including taking youth out on the land with elders to pass on the language and traditional knowledge and teachings, combining the transmission of traditional knowledge and wisdom with the creation of new economic opportunities for youth, and involving youth in the documentation and utilization of elders' knowledge in the school curriculum and in cultural and ecological restoration projects.

Examples include the Gwich'in project in northern Canada (14), the Whitefeather Forest project in Ontario, Canada (29) and the Jaru project in the Kimberley, Australia (16), all of which show that youth can become active participants in traditional and local ways when engaged and instructed by elders in their communities. The Apache students of Cibecue, Arizona (39) represent a case in which youth have cultivated a deeper cultural knowledge and a greater willingness to speak and develop proficiency in the Apache language. This has come after conducting interviews with elders in their community to better understand why their ancestors held such respect for water and reverence for sacred places.

Fostering the sustainable use and management of biodiversity for sustainable livelihoods

Several projects emphasize that traditional use and management practices are a means to maintain and conserve local biodiversity. Several other projects find ways to combine small economic development enterprises with biodiversity conservation. Examples of the former, where customary uses are encouraged, include the continued use of traditional resources for cultural purposes in urban settings in South Africa (25) and the maintenance of diverse rice landraces in Nepal (17), which are perpetuating the cultural value of these uses. The revival of traditional ways of nurturing agrobiodiversity in Peru (15) shows that where there is pride in the diversity of produce for consumption, there is an incentive for conservation. The Marine Protected Areas project in the Solomon Islands (12) clearly shows that people must be able, and be enabled, to conserve their own resources for their own benefit. In this case, traditional marine tenure systems are revived to allow for people to continue fishing (both for subsistence and for market sales) by managing the resource sustainably. In this project, people are made aware of the benefits that will accrue to them from protecting the reef ecosystem, via recovering the sustainability of the local fishery.

Small economic development projects are another means to promote biodiversity conservation through sustainable use. The loss of the means of subsistence and self-sufficiency is a reality affecting many regions worldwide, which is a challenge in some of the projects. Projects that foster small-scale economic activities promote sustainable livelihoods at one and the same time as they promote conservation of the resource being used. An example is the bottle gourd project in Kenya (37), which encourages the many uses of *kitete* for income generation. In the Tado cultural ecology project (23), the collaborating NGO supports community-based ecotourism initiatives, in which visitors participate in making traditional crafts and in preparing traditional foods and medicines, all of which both benefits the community economically and boosts conservation enthusiasm among community members. The Whitefeather Forest Initiative in Canada (29) combines environmental stewardship with economic renewal strategies to help the youth of the Pikangikum First Nation to develop new resource uses.

Similarly, the CAIMAN project in Ecuador (8) promoted income-generating activities compatible with the local indigenous communities' socio-cultural and environmental context (such as ecotourism and handicrafts), thus helping ensure ecological and economic sustainability. One way that the crocodile rehabilitation project in the Philippines (9) tackles the significant economic difficulties faced by approximately 40,000 people in the region is by supporting the enforcement of legislation that helps communities fish sustainably – thereby ensuring that the indigenous communities work towards maintaining a sustainable food supply.

Using traditional environmental knowledge in conservation planning

From the on-the-ground implementation of conservation projects that affect indigenous and local communities, to the development and application of environmental legislation at national and state or provincial levels, many of the projects stress and exemplify the relevance and benefits – for both people and the environment – of applying local and

Figure 6.2 *Mapping northern boreal forests in the Whitefeather Forest Initiative in northern Ontario, Canada, as part of a land use strategy that combines environmental stewardship with economic renewal opportunities for Pikangikum youth*

Credit: Whitefeather Forest Initiative

traditional ecological knowledge to conservation planning. For example, the use of traditional knowledge in provincial land use planning in British Columbia, Canada (13); the incorporation of aboriginal traditional knowledge mandated by the national Species at Risk legislation in Canada (26); the development of the national People's Biodiversity Register in India (33); and community forest management institutions in Cameroon (4) show that local knowledge is needed in planning processes, or, in the case of the biodiversity register, that local knowledge and innovation is a national resource. These projects also show that local knowledge is 'scaled up' to the regional or national level.

Challenges to the use of traditional knowledge in conservation planning occur in certain cases. In some instances, such as in the bottle gourd project in Kenya (37) and the Ethnobiology Project in Aurukun, Australia (31), problems occasionally arise with the sharing of biodiversity-related knowledge. Since, in some cases, such knowledge is specialized or considered sacred, sharing is possible only as long as the knowledge holders are comfortable with making that information public. Further, since traditional knowledge often has economic value, knowledge sharing also becomes an issue of knowledge holders' rights. In other cases, such as with the project in Yunnan, China (22), insufficient attention is given to local indigenous knowledge because scientific knowledge is privileged instead. With the Irular in Tamil Nadu (32), where responsibility for resource conservation was taken away from the community and transferred to government departments, the state has discretionary powers over resources, leaving local communities unable to apply their ecological knowledge.

Sometimes, an already fragile path to the affirmation of traditional knowledge in planning processes can be disrupted by even a single, unforeseen event, such as in the Muisca community in Colombia (36). In this case, the biggest challenge came with the untimely death of the community leader, who was a key knowledge holder of Muisca thought and cosmovision. This affected both the project and the community as a whole, and almost caused the project to collapse. Despite these enormous setbacks, the community has shown exceptional resilience, with community members finding ways to revitalize their culture, for example, by organizing traditional ceremonies for themselves as well as for non-Muisca who appreciate their cultural revitalization efforts.

Establishing collaborative partnerships

A number of projects illustrate cases of indigenous communities securing the means to design, manage and implement their own projects. Some of these projects are initiated and led by elders in the communities; others are initiated by youth or through community-wide efforts. All of these are exemplary of endogenous efforts that, so far, have shown to be successful. One such case among many others is the project in East Nusa Tenggara, Indonesia (23), where the indigenous groups run the entire project themselves and Tado and Waerebo research associates administer research programmes in their own communities. The Gwich'in place names project in northern Canada (14) is another example of a community-initiated and community-led project, where the work is carried out by the Gwich'in Social and Cultural Institute, the cultural and heritage arm of the Gwich'in Tribal Council, in collaboration with Gwich'in communities in the land claim area.

Among the sourcebook projects, there are also numerous examples of effective collaborative relationships between indigenous or local communities and outsiders, which are also considered successful. The Sierra Tarahumara, Mexico project (6) is an example of how lengthy and complex the process of full collaboration can be. Field visits and meetings with Rarámuri elders, traditional authorities and community members took place over several years before the communities chose to invite collaboration with outsiders. The project has also had a positive role as a catalyst within the community, by facilitating a number of community discussions and reflections on issues facing the communities that would probably not have happened otherwise, as well as by fostering collaborative work among Rarámuri community members that people might not have engaged in. In the case of the Xingu Indigenous Park (24), the collaborative partnership between the Xingu peoples and the NGO that helped them develop a 'life plan' for the park moved one step further when the NGO devolved activities to some of the park's indigenous groups, after it was determined that they were able to manage their own affairs.

Box 6.1 The requirements for Fully Collaborative Projects: A Kenyan Perspective

Community-based projects must be tailored to specific local needs, contexts and cultures, thus posing a big challenge. Such collaboration often requires time and flexibility that is difficult in tightly planned grants and projects. The process of learning from the community takes time and expectations are often high. Intellectual property issues and language communication barriers may also be a hindrance. These issues require sensitivity to the different circumstances of local people and outsiders. Establishing good rapport with local people is the key to the success of a community-based project. A minimum period of time is therefore needed to interact with the community to learn from each other. Sharing information becomes difficult when we touch on realms that interest a few individuals or specialized bodies of knowledge that [only a few can] claim supremacy on or sole rights to. This is aggravated further if there is an economic value at stake, a good case being the use of medicinal plants.

Yasuyuki Morimoto, contributor, project 37

Within collaborative partnerships, the learning process for all involved takes time and commitment. It implies listening, showing respect and familiarizing oneself with different worldviews, beliefs and values, knowledge systems, behaviours and languages. Reflecting this extra time requirement for collaborative projects is the statement reported in Box 6.1, drawn from the Kenyan bottle gourd landraces project (37):

The extent to which the projects we surveyed stressed a collaborative, participatory approach was one of our key points of interest. As we noted in the previous chapter, a project's participatory approach appears to positively correlate with the attention given to traditional knowledge and languages, and with the project's sustainability over time. As we also indicated in Chapter 5, in the present context it is difficult for us to go beyond these general observations, since we did not set out to undertake a systematic comparison of the outcomes of top-down vs. bottom-up and participatory approaches. It will be of interest to follow the outcomes of the projects in this sourcebook, in order to acquire a deeper understanding of what factors contribute to the most effective and successful partnerships in integrated biocultural diversity conservation efforts.

Focusing on capacity building

A capacity-building approach characterizes most of the projects in this sourcebook, whether the projects were initiated and managed by indigenous and local communities themselves or developed by outsiders. In the latter case, projects with a strong participatory emphasis tend also to stress capacity building for project sustainability. In both instances, project participants realize that long-term effectiveness in biocultural diversity conservation requires strengthening a variety of skills in the community and in all project participants, so that, over time, community members are empowered to take cultural affirmation and biocultural diversity conservation activities in their own hands. The variety of skills that are the focus of capacity building in the projects include good governance and transparency in community-based organizations for forest management in Cameroon (4); institutional capacity for ecotourism ventures and capacity as programme administrators in East Nusa Tenggara (23); capacity building for community members to carry out ecological restoration and improve landscape and community health in Sierra Tarahumara (6); organizational capacity and leadership to reverse the loss of culture, recover traditional political institutions, territorial defence, health care, education, agricultural production and traditional medicine for the Ngäbe in Costa Rica (20); training in basic ethnocartographic and ethnogeographic data collection and processing for the Hotï and Eñepa in Venezuela (37); ensuring full integration into the development and implementation of work plans to enhance capacity for biodiversity conservation within indigenous federations in Ecuador (8) and in Brazil (24); and the capacity to develop alternatives for the sustainable management of indigenous territories in Brazil (28).

The very focus on capacity building also brings its own challenges. For some projects, this may mean that activities take longer to get off the ground while the relevant skills are being developed. This is the case for the 'Ethnocartography and Self-Demarcation of Indigenous Peoples' Lands in Venezuela' project (35), in which the main challenges in assisting Hotï and Eñepa communities with geo-referenced mapping of their territories for the purpose of land claims were mostly technical and logistical. The project design emphasized training and empowerment, so that local people could become expert map makers and map users. Since local people had no previous experience with mapping technology, several weeks of training and supervised practice were required for each group. In some other cases, such as with the 'Eco-cultural Health in the Sierra Tarahumara, Mexico' project (6), capacity building may take even longer, to the extent that the first (and lengthy) step may be to spark an 'awakening' in the community – the realization that there are problems of a scale far larger than can be dealt with at the individual level, and that require unprecedented collaboration among community members. For the Gwich'in project in the Northwest Territories, Canada (14), the problem is of a slightly different nature. Efforts to record and revitalize the Gwich'in language are a vital part of the work involved in applying traditional knowledge to land use planning. However, largely due to the decline in the number of fluent language speakers, it is difficult to locate skilled people who can adequately transcribe and translate the language. Continued efforts to bring together elders and youth on the land to promote and pass on the language and knowledge about the land and the culture may help to correct this deficit in the future.

Enlisting government support

The importance of local, state or national government support for project success is apparent in many of the projects, and so are the challenges that arise from a lack of such support. Successful results in the 'Crocodile Rehabilitation, Observance and Conservation' project in the Philippines (9) are partially attributed to government support, insofar as the Local Government Unit (LGU) of San Mariano in the Sierra Madre of Luzon has become an active partner in crocodile conservation. The LGU has declared the Philippine crocodile the flagship species of the municipality, enacted local ordinances that protect the crocodiles, and established the very first Philippine crocodile sanctuary in the country. In the Bamenda Highlands Forest Project (4), the national government of Cameroon was a partner with local communities in the collaborative project. In the bottle gourd project in Kenya (37), the government acknowledged the success of the project and awarded the local women's group a small piece of land for a community centre and shop as well as a trophy for the best community-based income-generating project in the country.

On the other hand, examples of a lack of government support for biocultural conservation are also common in the projects. In Ethiopia (43), failure of the local government bodies to protect its indigenous peoples and their religious practices has resulted in inadequate protection for sacred forests. For the Andamanese dictionary project (40), there was a lack of will on the part of the government of India, which did not come forward to facilitate the project. Both of these projects have, however, managed to carry on, despite the lack of support: in the Ethiopian project, there were some concerned government members who became helpful; in the Andamanese case, the numerous bureaucratic hurdles were overcome, and the project's goals have been successfully achieved.

In the crocodile rehabilitation project in the Philippines (9), weak governance in local village councils has made it difficult for them to actively enforce environmental legislation that protects the wetland resources on which the communities depend. In the state of Acre, in the western Brazilian Amazon (28), over the past 35 years the forests have been adversely affected by large-scale Brazilian economic interests. These interests are backed by financial resources obtained from credit institutions and by Brazilian government incentives for the establishment of large cattle ranches, the exploitation of hardwood and agricultural activity. These incentives have led to considerable concentrations of private property, and serious conflicts have resulted from land takeovers, which have provoked confrontations between the 'new owners of Acre' and the local indigenous populations and rubber extractors. Gradually, as the indigenous peoples of Acre gain expertise in resource management skills, they will be more culturally resilient, economically active and more ecologically sustainable on their lands, which should help them in their efforts to protect their territories. In the case of the Muisca of Colombia (36), local authorities do not support the local community movement. At the national level, the Colombian government has abstained on the United Nations Declaration on the Rights of Indigenous Peoples. Government inaction has also hampered the community mapping project in Venezuela (35). The project's overarching goal was to assist two indigenous groups in producing the extensive documentation necessary for them to win legal title over their lands, but this goal has yet to be realized. Even though all of the required legal documents were assembled and handed in to the designated authorities several years ago, the government has made no decision on this case (or on any other land claims submitted by indigenous groups in Venezuela). Some observers in Venezuela doubt the government's willingness to live up to its commitment. Time will tell how the government will respond to the increased abilities of the Hotï and Eñepa peoples to demarcate and protect their traditional territories.

Concluding reflections on lessons learned

From environmental degradation to land use conversion, changes in biodiversity, over-exploitation of natural resources, economic development, land and resource tenure security issues and multiple forms of acculturation and forced social and economic change (see Table 5.1), the projects point to numerous major factors of socio-cultural and ecological transformation that are affecting biocultural diversity. As we discussed above, the projects also show that there are several conditions that are required to ensure success for biocultural diversity efforts in the face of these local and global challenges: strengthening institutions, ensuring land and resource tenure security, strengthening cultural identity, reconnecting elders with youth, sustainable use of biodiversity, use of local and traditional ecological knowledge in management, collaborative partnerships, capacity building and government support.

While acknowledging these conditions for success as significant steps to conserve biodiversity and support cultural resilience, it is also important to recognize that countering many of the factors that are negatively affecting biocultural diversity is often well beyond the control of indigenous peoples and local communities worldwide. These groups are, in many cases, confronting an uphill battle for self-determination and resource access. Providing the means to ensure conditions for success and at the same time to address the challenges calls for policy changes at the national and international levels.

While fewer in number and scope, there are other processes of change occurring at the local level that may be more within the control of local communities. Among these are, for example, deteriorating intergenerational (elder to youth) relationships, lack of institutional capacity and the decline in cultural identity, traditional teachings, and local languages. These recognizable impacts may prompt a call to local comunities to take stock and adopt the necessary measures to counter the underlying negative pressures. At the same time, given the processes of change that are impacting biological diversity and cultural resilience, the projects presented in this sourcebook illustrate clearly that biodiversity conservation and cultural affirmation cannot occur in isolation or at one scale only: revitalizing one requires similar efforts to recover the other, and conditions and effects at multiple scales must be taken into account. As a participant in the Philippine crocodile conservation project put it: 'There is a tendency in biodiversity conservation activities to focus on the mega scale and ignore the difficulties at the micro level. Only by addressing issues at the local level ("place"), and thus effectively making links with local livelihoods, cultural practices and beliefs, can we protect species' (Jan van der Ploeg, project 9). These remarks point to the need for greater efforts to understand what is happening at the local community level and how the issues link across scales, from the local to the national and international. They also call for national governments and international governing bodies to be accountable for the loss of diversity in all its forms at the local level, to listen to the voices of those who are working at the local level, and to understand and implement what needs to be done at national and global scales to help prevent further losses.

Most projects provide practical local-level examples of integrated biodiversity conservation and cultural affirmation – which in and of themselves are an excellent source of data and information for policy development to deal with loss of diversity.

It is one of the aims of this sourcebook to provide visibility to these projects so that policy development can learn and draw from these examples. Several contributors to this sourcebook stressed the need for increasing awareness of these community-based biocultural conservation initiatives and for placing the intimate relationship between cultures and biodiversity up front on the political agenda of conservation and development strategies. This points to a crucial gap that still needs to be filled and to the importance of 'connecting the dots' among these and many other similar efforts that are currently underway worldwide. These and other relevant points are addressed more specifically in Chapter 7.

The very diversity of the approaches taken by the projects, united by a common goal, is what makes for the projects' collective strength: each one of them exemplifies and addresses aspects of a whole constellation of issues that are critical for the achievement of biocultural diversity conservation and global sustainability. Looking at the projects as a kaleidoscope of human ingenuity put to the service of confronting some of the most pressing challenges of our times serves to highlight their very diversity as the key feature, instead of singling out individual projects as examples of 'best practices', as outlined in the methodology in Chapter 3. It is the collective dimension of these projects as a whole, rather than the features of any one 'model project', that reveals the variety and richness of '*good* practices' that are and can be deployed according to need and circumstances. Undoubtedly, a larger survey would have revealed an even greater pool of creative approaches and solutions.

It is clear that, while the sourcebook projects and many others like them around the world are making a difference at the local (and in some cases national and regional) level, considerable obstacles, as discussed above, remain in the way of promoting biocultural diversity conservation at all scales. In light of this, it is apparent that there is a critical need for awareness raising about indigenous peoples' rights and social, cultural and linguistic policies among all of those involved in biodiversity conservation, natural resource management and land use decisions. Virtually all project contributors argued that conservation cannot be done effectively using single-sector approaches and policies. An important lesson that can be drawn from the projects is that rapid social and ecological change is best addressed by the integration of approaches across disciplines and among governments, NGOs and intergovernmental organizations (IGOs) at the local, national and global levels. Major national and global efforts are also needed to remove the many barriers of state-controlled systems, the multiple causes of environmental degradation and the neglect of cultural diversity.

When looking at the current picture of the state of the global environment and the world's cultures and languages, the prospects for sustaining the biocultural diversity of life may not seem encouraging. Yet, the projects that make up this sourcebook provide solid evidence that there is a positive move afoot in terms of what is being done on the ground for integrated conservation. A long-term perspective is now needed to determine how these early efforts to integrate the strengthening of cultural resilience with biodiversity conservation are maintained by the communities involved, and the extent of the impact these and similar efforts will have. Currently, we are witnessing a surge of projects that are actively engaged in building this alternative. Addressing the many challenges will also depend on an increasingly favourable climate of acceptance of this kind of approach. By illustrating the nature and accomplishments of these projects, this sourcebook seeks to contribute to the creation of this climate. From a biocultural perspective, what is of central importance in the projects presented here is that they all address some of the key interconnections that confer vitality and resilience to social-ecological systems, and seek to maintain and strengthen them for the benefit of both cultural and biological diversity. Time will tell what contribution each of these projects will make to the enduring resilience of cultures and ecosystems in their respective locales and beyond – but they all stand out as examples of a new, integrated approach to sustainability that engenders hope for the future. In the next chapter, based on the analysis of the projects in Chapter 5 and the lessons outlined here, we offer a set of recommendations for strengthening and promoting a biocultural approach to conservation. The specific contribution of the biocultural approach to the future of sustainability is then addressed in the concluding chapter.

Part III

Sustaining Biocultural Diversity: Future Directions

7

Filling the Gaps and Connecting the Dots: Recommendations and Next Steps

Luisa Maffi and Ellen Woodley

The projects reviewed in this sourcebook teach us that fostering the biocultural diversity approach and its agenda for the sustainability of all life requires addressing challenges and opportunities at multiple levels. Numerous gaps need to be filled and connections need to be made in terms of research, policy and practice. Some of these needs are discussed below. But, above all, achieving biocultural sustainability requires change at a deeper level: a deep mind shift in what we value and cherish the most. In the concluding section of this chapter, we reflect on what such a shift entails.

Gaps and needs in biocultural research and documentation

When a new scientific concept with potentially far-reaching implications comes to the fore, it is not uncommon for it to encounter initial resistance, and to hear statements to the effect that 'more research is needed' to probe the concept. This has certainly been the case with the idea that the diversity of life is *biocultural* diversity, and therefore that sustaining the diversity of life means protecting and supporting both biodiversity and cultural diversity (including linguistic diversity), as well as the links between them. As an emergent and complex idea, it is intuitively graspable and appealing to many in an abstract way. Up to now, however, it has been more difficult to characterize its concrete, 'real-life' manifestations – the actual ways in which the 'biocultural nexus' between people and the environment establishes and perpetuates itself, can be disrupted and broken, and can be sustained or restored. This is at least one important reason why the theory and practice of biocultural diversity have not yet become mainstream, although the interest around them has been steadily growing. One key goal of this sourcebook has been to provide concrete data, analysis and lessons learned from on-the-ground biocultural projects, in order to shed more light and offer more insights on this idea and how it can be applied to sustain both cultures and biodiversity.

Inevitably, such an exercise has also unearthed a number of gaps in our knowledge and understanding of the on-the-ground dynamics of biocultural diversity, and of how attention to these dynamics can aid efforts at cultural affirmation and biodiversity conservation. Addressing such gaps is a priority for the further development of the field of biocultural diversity and its applications. Based on our review of the projects in this sourcebook, we identified the following gaps in documentation:

- Communities and researchers working on integrated biocultural conservation projects should generate more information on how strengthening local cultural knowledge, practices and languages can benefit the biodiversity and ecological health of the respective areas. Biocultural landscapes, protected sacred sites, protected habitat for culturally important species, community-conserved areas and conservation of diverse landraces through cultural traditions are examples of ways in which cultural values and the behaviours that stem from them contribute to maintaining high biodiversity and ecological values. More needs to be known about these and other cases of life-enhancing interactions between people and the environment.
- Similarly, more detailed information is needed on the biocultural values of 'place', to show how fostering and restoring ancestral connections to the land, and to the human history inscribed on it, can contribute to strengthening and reviving a sense of pride in and stewardship of local biocultural heritage, and thus to protecting and preserving that legacy. Among other benefits, this should help pinpoint the limitations of approaches that advocate for the exclusion of indigenous peoples and local communities from biodiverse regions and protected areas.
- Issues that have not yet been adequately addressed in biocultural research and practice – such as the role of gender in biocultural diversity conservation, as well as the biocultural values, knowledge and practices of urban indigenous peoples and non-indigenous rural or urban communities – should be explored and documented in greater depth. Elucidating these issues will significantly contribute to a more fine-grained and multifaceted understanding of the dimensions of biocultural diversity.
- A longitudinal documentation of the progress, changes, setbacks and successes that projects undergo (a task that, as we previously mentioned, was outside the scope of this sourcebook) is very important. Longitudinal studies can provide critical insights into the development of better and more efficient biocultural conservation practices and into what makes for project effectiveness and sustainability.
- Indicators that measure both the ecological and the cultural benefits of integrated biocultural conservation should be developed in order to provide evidence of the effectiveness of the biocultural approach. In order to measure the benefits ('success') of locally based biocultural diversity projects, such indicators should be developed jointly by project participants to ensure that locally meaningful parameters are duly taken into account.
- Using appropriate indicators, systematic comparisons are needed between the outcomes of projects that take an integrated biocultural approach and projects that do

not. Likewise, more research is desirable that specifically distinguishes between, and compares the outcomes of, community-initiated biocultural diversity conservation projects vs. those initiated by external researchers and agencies. Analysis along these lines would provide clarity on some of the conditions that may affect the success and long-term continuity of project activities, and address issues of community empowerment and self-sustainability.

In addition to further documentation, directions for future research include the following:

- Further efforts are needed to identify causal links between effective conservation and the maintenance of traditional and local values, beliefs, institutions, knowledge, practices and languages, in order to provide stronger guidance to conservation action. This requires systematically compiling existing knowledge on the cultural basis of conservation and promoting more interdisciplinary research on this topic. This research should be designed specifically to illuminate the connections between language, traditional knowledge and the environment, and to show how the persistence or erosion of one affects the others.
- Methods and tools for researching, measuring and monitoring the links between biodiversity and cultural diversity should be further developed. Current work on mapping and analysing the overlaps in the geographic distribution of biodiversity and cultural diversity at global and regional levels (described in Chapter 1) needs to develop into a full-fledged line of research, and to establish mutually enriching connections with local-scale mapping efforts (such as participatory mapping and ethnocartography). More work is also needed to devise and apply integrated biocultural indicators (cf. Chapter 1), in order to better assess the state and trends of biocultural diversity at global and regional scales, while at the local level communities can better monitor their cultural resilience and the vitality of their connections to the environment.
- Studies of the contributions of biocultural conservation projects to sustainable livelihoods may provide a more complete understanding of the relationship between environment, culture and poverty reduction, in the context of international processes such as the Millennium Development Goals. Such studies would help shed light on subsistence activities that can be both ecologically and economically sustainable, as well as compatible with the socio-cultural and environmental setting of indigenous and local communities. This research may also assist in mainstreaming the concept of integrating cultural affirmation with biodiversity conservation among the various international agencies concerned with environmental, social, cultural and economic sustainability.
- Greater support is needed for these various forms of integrated research and documentation that can advance our understanding of biocultural diversity and promote its practical application in policy and on the ground. To accomplish this, academic, funding, and other national and international institutions should strive to overcome the traditional disciplinary and sectoral boundaries that separate natural science research from social science research, environmental programmes from social programmes, and funding for biodiversity conservation from funding for

Figure 7.1 *Traditional handcrafts and sustainable livelihoods: A Waorani woman weaving a traditional bag using fibre from the chambira palm*

Credit: João de Queiroz

cultural (including linguistic) heritage and human development. Cross-disciplinary training that counters this fragmentation should be strongly promoted, so that the academics, practitioners and officers of relevant institutions will be more attuned to an integrated perspective towards sustaining cultures and biodiversity.

Filling these and other research and documentation gaps, while requiring major efforts, is critical to help advance an understanding of the nature and dynamics of biocultural diversity in both academic and professional circles, and promote the application of a biocultural approach in policy and on-the-ground work. At the same time, the need for more research in the future should not be a deterrent for taking action in the present, but rather, a precautionary approach should be taken. The precautionary principle states that, where there are threats of serious or irreversible damage, lack of full scientific certainty should not be used as a reason for deferring measures to prevent such damage. This principle has been invoked most often in the case of environmental issues, but it is applicable as well to social issues and the interaction between the two, as in the case of biocultural diversity. There is enough evidence already that biocultural diversity is being rapidly eroded, and that this poses serious threats for the vitality of our planet. More research will undoubtedly help refine our action plans to sustain and enhance biocultural diversity, but what we already know about the erosion of biocultural diversity should be enough to spur us to take action now.

Toward biocultural diversity policy: Advances and gaps

While still not at the forefront of policy and implementation, the biocultural diversity approach is acquiring increasing relevance, as conservation organizations, international agencies and, in some cases, national governments begin to include in their programmes and directives a concern for cultural diversity along with that for biodiversity.

At the international level, recognition of the importance of culture and cultural diversity for the conservation of biodiversity and for sustainable development was made explicit during the 2002 World Summit on Sustainable Development (WSSD) (UNEP and UNESCO, 2003). Both the Johannesburg Declaration on Sustainable Development and the Johannesburg Plan of Implementation issued by the WSSD call for respecting cultural diversity as essential for achieving sustainable development. This recognition is further reflected in guiding United Nations documents such as the Millennium Declaration, issued in 2000, which affirms the importance of the diversity of belief, culture and language and asserts that societal differences should be cherished as precious assets of humanity.

Some of the policies and programmes of the United Nations Environment Programme (UNEP), the United Nations Educational, Scientific and Cultural Organization (UNESCO), the Convention on Biological Diversity (CBD), the International Union for the Conservation of Nature (IUCN), and other international organizations now acknowledge the interrelationships between biodiversity and cultural diversity. Highlights of these policies and programmes are shown in Box 7.1.

Some national policies have also taken the initiative to strengthen the links between biological and cultural diversity, especially in compliance with the CBD. For example:

- The Biological Diversity Act of India (2002) stipulates that the central government shall endeavour to respect and protect the knowledge of local people relating to biological diversity. Forests that are protected as sacred groves, based on local communities' belief systems, may be recognized as heritage sites under the Act.
- In the Philippines, the government passed the Indigenous People's Rights Act in 1997 that explicitly recognizes the rights of indigenous peoples to their ancestral lands, to self-determination and to the free exercise of their culture. Around 76,000 indigenous people (out of the total indigenous population of 8 million) are direct beneficiaries of Certificates of Ancestral Domain, which recognize their inherent right to self-governance and self-determination, and ensure respect for the integrity of their values, practices and institutions (UNDP, 2004).
- The Republic of Panama legally recognizes the sovereignty of seven indigenous groups. Panama was the first government in Latin America to recognize this class of rights for indigenous populations, and now 22 per cent of the national territory is designated as sovereign indigenous reserves (Condit et al, 2001).
- Several other Latin American countries, such as Colombia, Venezuela and Ecuador, have variously given recognition to the land rights and cultural rights of the local indigenous peoples and passed land demarcation laws, although implementation often remains problematic.

Box 7.1 International Policies and Programmes that Support Biocultural Diversity

- UNEP complemented its Global Biodiversity Assessment (Heywood, 1995) with an extensive review of the cultural and spiritual values of biodiversity (Posey, 1999), including the role of linguistic diversity (Maffi et al, 1999).
- UNEP's GEO-4 report (2007) defines biodiversity as including also 'human cultural diversity, which can be affected by the same drivers as biodiversity, and which has impacts on the diversity of genes, other species, and ecosystems.'
- An initiative on science and traditional knowledge was carried out by the International Council for Science (ICSU, 2002), following up on some of the outcomes of the UNESCO World Conference on Science (UNESCO, 2000).
- UNESCO adopted the Universal Declaration on Cultural Diversity in 2001 and the Convention on the Protection and Promotion of the Diversity of Cultural Expressions in 2005.
- UNESCO's Endangered Languages Programme aims to safeguard the world's linguistic heritage, while its LINKS (Local and Indigenous Knowledge Systems in a Global Society) programme focuses on the strengthening and revitalization of traditional knowledge.
- UNESCO also has a Main Line of Action on Biodiversity and Cultural Diversity and organized two interdisciplinary expert meetings on the theme of linkages between cultural and biological diversity in Aichi, Japan (April 2005) and Paris, France (September 2007). These meetings yielded the publications *Learning and Knowing in Indigenous Societies Today* (UNESCO, 2009) and *Links Between Biological and Cultural Diversity* (UNESCO, 2008), respectively.
- UNESCO's Programme on Man and the Biosphere (MAB) recognizes that traditional forms of land use often conserve ancient breeds of livestock and crop landraces.
- Article 8j of the CBD explicitly acknowledges the important contribution of traditional knowledge to the conservation and sustainable use of biological diversity.
- The CBD's 2010 Target (which aims to significantly reduce the loss of biodiversity by the year 2010) includes as one of its focal areas (Focal Area 5) assessing the status and trends of indigenous and local knowledge relevant to the conservation of biodiversity, for which status and trends of linguistic diversity has been chosen as a proxy (given the current lack of global quantitative data on TEK).
- The Akwé: Kon Voluntary Guidelines developed by the CBD centre on impact assessment procedures and methodologies. Through the cultural impact assessment process, cultural issues to be considered are 'cultural heritage, religions, beliefs and sacred teachings, customary practices, forms of social organization, systems of natural resource use, including patterns of land use, places of cultural significance, economic valuation of cultural resources, sacred sites, ceremonies, languages, customary law systems, and political structures, roles and customs'.

- An initiative by the United Nations University on traditional knowledge proposes to respond to questions related to how traditional knowledge is being considered in intergovernmental processes related to environmental conservation, sustainable development, human rights, international trade and intellectual property (United Nations University, 2005).
- A UN-led initiative aiming to help protect sacred sites worldwide as places where spiritual values have contributed to the conservation of biodiversity was launched at the CBD's Eighth Conference of the Parties (COP 8) in 2006.
- The UN Declaration on the Rights of Indigenous Peoples, approved in 2007, states that 'control by indigenous peoples over developments affecting them and their lands, territories and resources will enable them to maintain and strengthen their institutions, cultures and traditions, and to promote their development in accordance with their aspirations and needs'. It also recognizes that 'respect for indigenous knowledge, cultures and traditional practices contributes to sustainable and equitable development and proper management of the environment' (United Nations, 2007).
- IUCN's Fourth World Conservation Congress (WCC), held in Barcelona, Spain in October 2008, had as its theme 'A Diverse and Sustainable World' and included a week-long 'Biocultural Diversity and Indigenous Peoples Journey' during its Conservation Forum. IUCN's programme of work for 2009–2012, approved at the WCC, recognizes the importance of cultural values as related to nature and of cultural diversity as 'an important safeguard for both ecosystems and social systems', and identifies the cultures of indigenous and traditional peoples as 'vivid examples of the profound and lasting connections between cultural and biological diversity'.
- IUCN's Commission on Environmental, Economic and Social Policy (CEESP) includes in its vision that of 'a world where cultural diversity is intertwined with biological diversity', and identifies the 'three-pronged crisis of energy, climate change and biocultural diversity loss' as the great challenge of our times. Issues of culture and conservation were designated as one of CEESP's priorities in its 2005–2008 programme of work, with the objective of 'improved knowledge, policy and practice linking cultural and biological diversity, distancing their common threats and strengthening their common opportunities'. CEESP's mandate for 2009–2012 continues this focus, having the promotion of biocultural diversity and reducing impacts on it among its key objectives.
- IUCN/CEESP also established a Theme on Culture and Conservation (TCC) for the specific purpose of supporting the development of international policies that are sensitive to the cultural dimensions of biodiversity and of nature conservation. In CEESP's 2009–2012 mandate, TCC's objective is 'improved knowledge, policy and practice linking biological diversity and the cultural dimensions of nature conservation, reversal of the loss of biocultural diversity, and promotion of socio-environmental wellbeing'. TCC focuses 'on the conservation of biocultural diversity through improved understanding, applied research and policy advice on the relationships between culture and biodiversity conservation'.
- The UN Food and Agriculture Organization (FAO), in collaboration with the International Indian Treaty Council (IITC), sponsored work on the development of indicators of the rights to food and food sovereignty for indigenous peoples, which recognizes the importance of culture in sustainable development.

All of these and other related developments are positive signs of change toward greater recognition of the relevance of cultural diversity for biodiversity conservation and environmental and social sustainability. Indigenous peoples' and local communities' organizations have been actively involved in many of these processes, influencing outcomes at international and national scales. In response to the multiple external pressures to which their communities are exposed, their main focus at the political level has been on the development and affirmation of their rights. This has included in particular land and traditional resource rights; intellectual property rights over traditional knowledge relevant to the conservation and sustainable use of biodiversity; and access and benefit sharing rights related to resource extraction from their lands and territories (The World Rainforest Movement, 1992; Tebtebba Foundation, 2002; Human Rights Council, 2006; IIFB, 2006; UNPFII, 2006).

These and other advances at the international and national levels in the development of policies supportive of biocultural diversity stem from the growing recognition, over the past two decades, of the importance of culture and cultural diversity in relation to the environment and human well-being. There remain, however, significant policy gaps to be filled, including the following:

- It is necessary to fully acknowledge the role of culture in sustainable development. In its original formulation, the sustainable development paradigm (Brundtland Report, Agenda 21) does not explicitly address issues of culture. The cultural approach to development is a newly emerging strategy that is beginning to recognize the importance of place-based knowledge, beliefs and practices for sustainable development. The Sustainable Livelihoods Framework was developed by the UK Department for International Development in 1990 and has subsequently been used and adapted by various development agencies around the world. This framework could be modified so that culture is more explicitly recognized as cutting across all of the five livelihoods assets (human, financial, physical, natural and social) (Woodley et al, 2008). Many development interventions in the past considered certain cultural practices, such as customary land tenure systems and traditional knowledge, to be 'impediments to development' (Riddell, 2000; Stavenhagen, 2000). While the tide is turning, much remains to be done on the way to achieving a redefinition of 'development' – and thus of sustainable development – from an endogenous biocultural perspective: a perspective in which cultural groups and communities are empowered to establish their own definitions of and paths to development from within, in harmony with their cultural and biological heritage (Maffi, 2007b). Box 7.2 presents an alternative, endogenous definition of development that was formulated by the indigenous participants in the 2nd Global Consultation on the Right to Food and Food Security for Indigenous Peoples, held in Nicaragua in 2006.
- There is a need for increased awareness of, and for acting upon, the social, economic and environmental impacts that the forces of globalization are having on some of the most vulnerable sectors of the world's population, such as indigenous and local communities. All human societies are ultimately susceptible to these impacts, but the most immediately vulnerable are those groups that rely directly on local natural resources for their livelihoods and well-being. These people are seeing their resource base being rapidly depleted and their social and economic structures being severely

Box 7.2 An Indigenous Peoples' Definition of Development

Development with identity is the project of life of the Indigenous Peoples based on their own logic and worldview. It is the natural growth of Indigenous Peoples, of their flora and of their fauna based on principles of self-determination in relation to land, territories, and natural resources. It is also respect for their individual and collective rights. It is the welfare and security of our peoples.

Woodley et al, 2008

undermined. Countering these impacts calls for policies that address mitigation, coping and adaptive capacity of these vulnerable groups, so as to provide them with greater control over the forces of change.

- Greater efforts are also needed to support and strengthen these groups' institutions and learning processes, so as to ensure that they can be fully informed about and effectively involved in conservation and development choices and decisions that may have long-term implications for their well-being and the health of the ecosystems in which they live. Policy that is respectful and inclusive of cultural values and traditions should both support and encourage the development of community-level institutions in order to avoid top-down approaches and promote the scaling up of local-level skills, practices and institutions (Ericksen and Woodley, 2005).
- To ensure greater communication and more effective partnerships between vulnerable people and conservation policy makers, conservation practitioners must take on the responsibilities of a human rights approach to conservation and join civil society efforts to create more socially just societies (Alcorn and Royo, 2007). Indigenous peoples' rights to traditional lands and resources can no longer be ignored. There is an urgent need for genuine collaboration to protect biodiversity while respecting the rights, needs and aspirations of the traditional stewards of areas that are the object of conservation efforts.
- There is a need to promote more in-depth understanding of the importance of supporting local and traditional knowledge and languages for the maintenance of both biological and cultural diversity, and thus for sustainability. Not enough substantive attention is being devoted as yet to the role of local and traditional knowledge in the formulation of conservation and development policies and in on-the-ground action, in spite of the statements of principle contained in international documents and initiatives such as those mentioned in Box 7.1. Further progress will necessitate mainstreaming the interlinkages between biodiversity and culture into social and environmental plans and policies, thus working to strengthen the interface between policy and traditional science.
- An additional policy gap is the need to move from the current focus on maintaining and protecting traditional and local ecological knowledge as a valuable aspect of human heritage to a focus on strengthening the vitality and intergenerational transmission

of this knowledge and of the languages through which knowledge is transmitted. This requires identifying and supporting the formal and informal institutions and practices that are involved in this transmission. Language and culture affirmation and revitalization activities are taking place in indigenous and local communities worldwide, and need to be promoted and made more visible. However, many governments are still reluctant to support and promote mother-tongue education and a culturally relevant curriculum for indigenous peoples and minorities – mostly for political reasons, although the arguments are often couched in economic terms. Further development of the emerging field that studies the economic and social costs of loss of language and culture vs. the cost of supporting their maintenance (e.g. Grin, 2005) should help dispel prevailing myths in this regard, such as that bilingual or multilingual and multicultural education is 'too costly'. In fact, it only appears to be so when the real costs of cultural dislocation and forced assimilation are externalized and not computed in these calculations. In another relevant development, economic theory is beginning to address the significance of culture as the interface between humans and nature, and of 'cultural capital' as the interface between natural capital and human-made capital (Cochrane, 2006; also see Berkes and Folke, 1994). A better understanding of the social, environmental, as well as economic values of both biodiversity and cultural (including linguistic) diversity will help to further advance an integrative approach to biodiversity conservation and cultural affirmation for sustainable development.

The need for enlightened environmental and social policies has never been greater and more urgent, as biocultural diversity continues to decline globally, despite the growing recognition of its vital importance for the future of our planet. Governments and international organizations should fully recognize this predicament and its implications for humanity and all life on Earth, and take the lead in forging a new integrated biocultural path to sustainability.

Promoting a 'community of practice' in biocultural diversity conservation

One important way to promote a new path to sustainability is to support on-the-ground efforts that integrate biodiversity conservation and cultural affirmation. As we pointed out in the Introduction, efforts of this nature abound worldwide, but so far have tended to be carried out in isolation, with no established mechanisms for making the interconnections. This has limited their ability to gain global visibility and make a mark beyond the local level. One of the goals of our sourcebook project has been to create the conditions for greater direct interactions among researchers and practitioners involved in biocultural conservation activities. Making the connections among the projects reviewed here, as well as others, supports the development of a network, or 'community of practice', in biocultural diversity conservation. Within such a network, people are beginning to share information, experiences and lessons learned among peers, and to build on this knowledge sharing in order to strengthen methodologies, expand the scope of the approach, and identify needs and opportunities for promoting biocultural diversity research and action, and increasing the visibility of these efforts.

Several sourcebook contributors commented on the importance of developing a biocultural diversity network. Their comments expressed the need to share with one another what works and what remains a challenge, as well as to generate greater awareness of the importance of integrating cultural values, knowledge and practices with biodiversity conservation. Contributors felt that developing a community of practice in biocultural diversity conservation will achieve several goals, in terms of opportunities for sharing ideas and experiences and for creating greater visibility for integrated conservation approaches. In their opinion, a network will:

- enable people to identify and contact others who have similar concerns, in order to discuss theoretical assumptions, on-the-ground work, problems and successes; establish partnerships; and learn from one another through sharing experiences from different points of view and from different local contexts;
- help elders, community members and project participants disseminate and discuss information about 'good practices', and what works and what does not in specific contexts, thus potentially saving time, energy and resources;
- broaden our collective understanding of what contributes to successful alliances between indigenous peoples and those involved in conservation efforts, extractive industries and other activities taking place in indigenous territories.

As for how a community of practice can increase the visibility of integrated biocultural conservation projects, sourcebook contributors indicated that it could do so by:

- raising national and international awareness of the important role that indigenous peoples and local communities play in biodiversity conservation, and of the difficulties they face in maintaining their traditional control over and management of their lands and territories and their natural resources;
- promoting more culturally sensitive and politically responsible behaviour among outsiders operating in indigenous territories;
- giving greater recognition to the role of community-based resource management and the relationship between cultural affirmation and biodiversity conservation;
- increasing access to funding opportunities for similar projects and helping governments see the need to provide funding for aboriginal language education and aboriginal curriculum rooted in the land and in local cultural traditions;
- spreading awareness among policy makers and the general public about the issues of language and cultural loss and about the links between the loss of biodiversity and the loss of cultural values, beliefs, institutions, knowledge, practices and languages, as well as emphasizing the importance that biocultural diversity holds for the people involved;
- helping place biocultural diversity conservation on the political agenda of national development strategies as a human rights issue;
- providing a forum to encourage international cooperation concerning biocultural diversity issues.

In response to the interest expressed by sourcebook contributors, as we mentioned in the Introduction, we have taken an initial step toward building a community of practice in biocultural diversity conservation through the creation of a companion portal to this sourcebook (www.terralingua.org/bcdconservation). On this portal, an interactive database of biocultural diversity conservation projects allows for new contributions to be added, thus progressively expanding the network and its worldwide reach. Also on the portal, an electronic discussion forum enables participants to post queries and comments and discuss relevant topics, ranging from the 'nuts and bolts' of biocultural diversity conservation, to planning common activities and strategies, to creating sub-networks among projects in the same region, and so forth. The forum provides participants with the means to learn from one another and develop strategies for strengthening their own projects as well as for becoming collectively more effective in pursuing shared goals related to fund raising, policy development and advocacy. We expect that an interconnected network of practitioners of biocultural diversity conservation will contribute to raising the visibility of this approach and to illuminating the relevance of the biocultural approach for sustainability. It will thus help create more auspicious conditions for the protection, maintenance and revitalization of biocultural diversity.

Synergizing with the conservation community

An active network of biocultural diversity conservation practitioners can also contribute valuable information to other relevant networks and discussions, such as those related to the co-management of protected and other ecologically sensitive areas from which the traditional inhabitants have been excluded, or to which they have limited access (Borrini-Feyerabend et al, 2004). In the conservation community, strict nature reserves, wilderness areas and National Parks, which are by definition exclusive of human habitation – or at least of permanent human habitation (IUCN categories I and II) – are generally considered necessary for ecosystem protection and critical to the global effort to fight rapid biodiversity loss. However, there is an urgent need to recognize the consequences – for both humans and the environment – of the resulting displacement of people who may have lived within those ecosystems for thousands of years, as well as of the loss of the associated place-based values, knowledge and practices. When the vital connection to 'place' is removed in the name of ecosystem protection, the maintenance of cultural identity and resilience is threatened, along with the traditional knowledge and practices associated with these ecosystems. How to handle such critical issues is best decided collectively by the local communities who are impacted; however, the issues also need to be acknowledged, and solutions supported, by the conservation community.

In co-management approaches, an awareness of the negative impacts that a disassociation with 'place' has on cultural identity and resilience, and consequently on environmental conservation as well, is the first step towards addressing cultural erosion under these circumstances. In such cases, it is not sufficient, although it is important, to engage in 'salvage' operations, such as documenting traditional values, knowledge and practices, and ensuring that local traditions and languages are taught in the school system so that a sense of pride in culture is maintained. When values, knowledge, practices and languages have become 'decontextualized' due to physical displacement of local people, 'salvage' initiatives cannot, by themselves, guarantee the continued vitality of such elements of culture. It is essential for the culturally based worldviews, knowledge and practices of the local people to be returned to their context and be fully integrated into conservation and management plans. Integrated efforts such as those represented by the projects in this sourcebook provide significant examples of how to effectively do so in the context of co-management arrangements.

Furthermore, the biocultural diversity conservation approach shares key characteristics with the concept and practice of Indigenous and Community Conserved Areas (ICCAs), which are gaining visibility worldwide. ICCAs are defined as 'natural and/or modified ecosystems containing significant biodiversity values, ecological services and cultural values, voluntarily conserved by indigenous and mobile peoples and local communities through customary laws or other effective means' (www.iucn.org/about/union/commissions/ceesp/topics/governance/icca/index.cfm). The Fifth World Parks Congress (WPC) of 2003 and the ensuing Programme of Work on Protected Areas of the CBD accepted ICCAs as legitimate conservation areas to be supported and, as appropriate, included in national and international protected area systems. Following WPC recommendations, the CBD called for the signatory countries to better understand and appreciate local knowledge and the priorities, practices and values of indigenous and local communities, and to identify and remove the barriers that prevent adequate participation of local and indigenous communities in all stages of protected area planning, establishment, governance and management. Some national governments have integrated ICCAs into their official Protected Area Systems.

The biocultural perspective adds specificity to ICCA debates, by explicitly pointing to one of the crucial requirements for sustaining ICCAs: the need to support the cultural values, beliefs, institutions, knowledge systems and languages that underlie the traditional institutions and practices related to conservation. This perspective highlights the fundamental link between biodiversity conservation and cultural affirmation at the community level, which needs to be taken into account and strengthened in on-the-ground work and policy. In turn, biocultural practitioners will gain from expanding their horizons through interactions with people involved in ICCA-related activities, in particular by learning more about the international policy context relevant to both kinds of efforts. This 'cross-fertilization' should encourage greater understanding and appreciation of the role of indigenous and local communities in conservation, for the benefit of both biological and cultural diversity.

A large network of biocultural diversity conservation practitioners may act as a catalyst in international discussions about the rights and contributions of indigenous peoples and local communities, who often have had little or no say when their traditional territories become the object of conservation efforts or extractive activities, or when they are set aside as protected areas. The debate continues, and in this context a biocultural network can provide valuable information and experience on how to successfully integrate cultural affirmation and biodiversity conservation.

The need for education and a shift in values

In spite of much progress and many encouraging signs, it is clear that in the world today such an integrated approach is still far from mainstream. Indeed, many powerful forces seem to continue to push us in the opposite direction. Education – not just as information, but education of the kind that deepens understanding and transforms moral and spiritual values – is what is ultimately required to produce this societal shift. This is the single greatest challenge for the sustainability of the diversity of life.

Figure 7.2 *Parents and children in Pitumarca, Cusco, Peru, posing next to a poster that compares traditional and Western forms of knowledge*

Credit: Jorge Ishizawa

This volume and the projects showcased in it represent a small but meaningful contribution in this direction, one positive step among others that are beginning to affect how people think and the choices they make. Among those choices are political choices that can move political will toward an increasingly integrated policy approach to protecting and enhancing the biocultural diversity of life. The signs of positive change in political decision making mentioned in this chapter, as well as others, are an indication of movement in the right direction. Yet, much more needs to be done to foster the kind of integrative thinking and action that will in turn create a societal climate favourable to biocultural diversity and help counter the many forces (reviewed in Chapter 5) that are negatively affecting it. This educational goal calls for filling several important gaps, some of which are outlined below.

- As we mentioned earlier in this chapter, the 'institutional culture' of many academic, non-governmental, governmental and international organizations, where fragmented sectoral approaches are still common, has represented an obstacle to the widespread adoption of an integrated biocultural approach. This shortcoming has resulted in placing a heavy burden on those other organizations – mostly small NGOs and community-based organizations – that have actively been seeking to overcome fragmentation and promote integration. Education campaigns are exceedingly labour-intensive and costly for these organizations. Larger organizations (including conservation organizations, which commonly see their central role as one of saving the world's natural heritage rather than being involved in 'ancillary' social and economic issues) need to educate themselves about the interdependence of biological and cultural diversity – and they are beginning to do so. These efforts should be fostered and expanded.

- Funding institutions at all levels (governmental, intergovernmental or private) also need to educate themselves about biocultural diversity. As mentioned above, many of these institutions still exhibit the kind of structural fragmentation in their programmes that hampers support for integrative, cross-disciplinary and cross-sectoral initiatives. In this case too, a handful of innovative and visionary funding institutions has taken on the challenge of promoting integration and supporting bioculturally oriented research and action. As awareness of the biocultural approach grows among other funding institutions, they will in turn be able to act as educators and promoters vis-à-vis other institutions, organizations, governments, media and the public at large.
- Ultimately, creating a more favourable climate for the acceptance and adoption of an integrated approach to biodiversity conservation and cultural affirmation, and for firmly enshrining the relationship between the two in national and international political agendas, is a matter of promoting a profound shift in understanding and values among the general public. It is apparent that large sectors of humanity have become deeply disconnected from the natural environment, and thus from the perception of our continued dependence on, and interdependence with, the ecosystems we live in. This disconnect tends to make people inured to the environmental consequences of our actions and to the ways in which those consequences in turn negatively affect human well-being. Under these circumstances, it is common to hear appeals to ever new technological 'fixes'. This reveals a profound lack of awareness of how social-ecological systems, once their health is radically compromised, may go beyond the point of no return – that is, beyond the point of repair by whatever technological means. To bring about this societal shift in understandings and values is undoubtedly the most significant and challenging of the educational efforts required.

What is necessary at all levels is a radical change of mind towards that 'logic of interconnectedness' that – as one sourcebook contributor puts it (see Chapter 6) – 'brings along with it concepts of integrity, responsible action and sound relationships, and the idea that all of our actions have consequences'. A key aspect of embracing this logic of interconnectedness is adopting the view that the diversity of life is diversity in both nature and culture, and that biodiversity and cultural diversity (including linguistic diversity) must all be cherished and cared for as an interlinked whole that is both the product of the evolution of life and the expression of its future potential. In the concluding chapter, we consider the implications of this view for the future of sustainability.

8

Biocultural Diversity and the Future of Sustainability

Luisa Maffi

In 1987, the Brundtland Report identified three components of sustainability – environment, society and economy – and defined sustainable development as 'development that meets the needs of the present without compromising the ability of future generations to meet their own needs' (WCED, 1987, p43). In broad strokes, this definition sought to harmonize the demands of socio-economic change with the need to ensure continued availability of the Earth's natural resources over time. By so doing, the report was still, by and large, taking an economics-driven perspective on the natural world – one that views nature principally as a store of valuable resources for human use rather than as an interconnected web of life and provider of life-sustaining functions for all living beings. At the same time, it did enshrine some fundamental realizations: that the planet's natural resource base is not infinite; that its replenishment capacity has not been keeping pace with the rhythm and scale of economic development; and that this has severe consequences for both the environment and people, particularly the more vulnerable sectors of society.

The report's dictum about sustainability has been a major source of inspiration for a vast movement that has been striving to realize a 'three-pillar' form of balance among environment, society and economy. As this sourcebook clearly shows, there is now ample evidence for the need to include culture as another fundamental dimension of sustainability. In other important respects, however, the key issue is not so much whether the model of sustainability should be a three-pillar or a four-pillar one – or even whether it should be represented in an altogether different fashion. Rather, it is a matter of reconsidering how the concept itself has fared in the intervening years since the Brundtland Report. Paradoxically, the dismal trends revealed by the most recent global assessments of the world's ecosystems and biodiversity, which we mentioned in the Introduction, make it apparent that, no matter how popular the concept has become, and in spite of a wealth of genuine efforts to realize its vision, its goals are far from having been achieved.

The idea of sustainability itself, while now invoked almost as a mantra in nearly every programme of work, seems in many ways to have become more vague and diffuse over time, having come to mean rather different things for different people and in different contexts, with its original objectives remaining by and large elusive (see e.g. Adams, 2006; Adams and Jeanrenaud, 2008). Looking at the 'big picture', it seems woefully evident that overall, if anything has been 'sustained' in the intervening years, it has been 'business as usual': economic growth, at the expense of both environment and society – and at the added expense of culture. Even with a recession as severe as the one the world experienced in 2008–2009, the balance has not shifted away from the imperative of economic growth and toward a recognition of ecological limits and of the economy as a part of the ecology, rather than the other way around.

In the context of 'business as usual', the social and cultural risks of unfettered economic growth have not substantially come into the picture any more than the ecological risks of it have. The very philosophy of economic growth is predicated on the assumption of limitless resources – in other words, on denial of the fact that we live within ecological limits. Nor does that philosophy recognize the interconnectedness and interdependence of ecological, social and cultural dimensions. Several powerful contributing factors have been at play, including the common (although by no means universal) human propensity for short-term thinking and immediate satisfaction of one's needs and desires, the belief that the Earth's ecosystems will always recover from the destructive pressures human demands place on them, and the faith in the 'technological fix', which (as we pointed out in the previous chapter) has lulled many into the false expectation that cleverly engineered solutions can always be found for whatever environmental problems we bring about.

Even the specialized discipline of ecological economics, which was created with the intent to address the neglect of environmental (and social) considerations in classical economics, has by and large continued to respond to the imperatives of classical economics. It has brought to the foreground the idea of 'valuing nature's services', and thus allowed for the mistaken and misleading – if conveniently reassuring – assumption that trade-offs of 'ecosystem services' for certain economic goals are generally possible (that is, possible in more than in very limited and very specific cases).

A more ground-breaking ecological economics should rather take as its foundation the recognition that ecological functions – life's support systems that are the very basis of the health of ecosystems and people and of the well-being of societies – are priceless and irreplaceable. On that basis, such a 'new' ecological economics would seek to forge a truly sustainable economy: an economy of ecological limits, not one of continued economic growth. Such a true, radical departure from classical economics, however, is yet to materialize.

Confronted with this state of affairs, many commentators have concluded that the notion of sustainable development is too loose, or has become too skewed, to be of much use. At the same time, others have suggested a strategy of 'keep it but fix it', by 'reorientating the concept of sustainability, re-emphasizing what it means and moving forwards' (Adams, 2006, p10). We would agree: what needs to be discarded is not the idea of sustainability; it is the state of denial that has not allowed for the true meaning of sustainability to emerge. From a biocultural point of view, it is quite clear what 'sustainability' means. Another commentator puts it eloquently: 'What is to be "sustained" in this rapidly changing world? The answer is simply yet profoundly "life itself"– life in its richness, diversity, vitality, and resilience in both nature and culture' (Rapport, 2008, p1).

It is increasingly recognized that resilience is a central concept in rethinking sustainability. What the biocultural perspective (along with other germane approaches, such as the study of the dynamics of social-ecological systems) makes clear is that when we refer to resilience we should be thinking not only of the ecological resilience of the biosphere (its capacity to rebound from stress and return to its former state once the stress is removed), but also of the cultural resilience of the 'ethnosphere' and the 'logosphere' – the planetary webs of peoples and languages (see Chapter 1) – and the resilience of the very interconnectedness among all three. Just as we now understand that ecosystems are not infinitely resilient to stress and, under continued pressure, can go into irreversible breakdown, we are beginning to understand that social systems are affected in similar ways and can be damaged beyond repair. As we saw in Chapter 5, nature and culture are in fact affected by many of the same factors that lead to loss of organization, vitality and resilience – from environmental degradation and over-exploitation of natural resources, to development pressures, to issues of land use and resource tenure, to the introduction of non-native species and monocultures, to acculturation and other socio-economic changes. The interconnection of human societies and ecosystems – mediated by cultural values, beliefs, knowledge, practices and languages – implies that a loss of resilience anywhere in an integrated biocultural system is likely to contribute to loss of resilience elsewhere. 'Sustaining life', therefore, comes to mean 'sustaining life in nature and culture' – the biocultural diversity of life.

This realization must now become central to the agenda of sustainability. And, indeed, there are notable signs of movement in that direction. Importantly, some of these signs come from IUCN, the world's largest and most influential conservation organization, endowed with a special ability to influence standards and practice. For instance, in a 2008 report written by IUCN's Director General in preparation for the Fourth World Conservation Congress, one finds the following statement:

> *We have ... achieved a better understanding of the nexus between the diversity of living beings and the diversity of cultures – which together make up the diversity of life on the planet. Nurturing human diversity through culture-based conservation, maintenance of traditional knowledge, revitalization of local practices of natural resource use and governance have become equally important objectives of IUCN as those of conserving species and ecosystems – because ultimately they are profoundly linked realities.* (Marton-Lefèvre, 2008, p49)

'Shaping a Sustainable Future', IUCN's programme of work for 2009–2012 (IUCN, 2008) further articulates this 'better understanding of the nexus between the diversity of living beings and the diversity of cultures', as seen in Box 8.1.

Box 8.1 Cultural and Ethical Values in IUCN's 2009–2012 Programme of Work

Cultural values and ethics are important foundations of human behaviour, particularly in relation to nature. In a globalized world that tends to homogenize cultures, cultural diversity provides an important safeguard for both ecosystems and social systems. It embodies the human experience of interacting with nature throughout history, civilizations and landscapes, and therefore represents the cumulative wisdom and skills of humanity to manage nature and natural resources.

The significant overlap seen in the world between biological and linguistic diversities, as exemplified in Oceania or Mesoamerica, is a case in point. This geographic overlap speaks of interlinked processes of diversification, resulting in thousands of different cultures living in diverse environments that they contribute to shape. The cultures of indigenous and traditional peoples are vivid examples of the profound and lasting connections between cultural and biological diversity.

Beyond traditional societies, cultural background and behaviour affect the drivers of biodiversity loss. These behaviours, and the resulting impact on biodiversity, can change, especially now that formal and informal networks for information exchange and learning have emerged worldwide on a range of issues, including on the valuation of nature and ecosystem services, sometimes leading to the designation of cultural land/seascapes.

IUCN, 2008, p18

Transition to Sustainability: Towards a Humane and Diverse World, a document prepared for IUCN's 'Future of Sustainability' process, echoes some of the same points:

> *There are remarkable parallels and linkages between the distribution and persistence of biodiversity and of cultural and linguistic diversity, and numerous case studies demonstrate that cultural diversity is integral to the conservation of landscapes and other aspects of biodiversity. We need a collaborative approach to retaining diversity on earth, not separate or conflicting strategies for dealing with the component diversities separately.* (Adams and Jeanrenaud, 2008, p55)

Similar sentiments are expressed in a recent analysis of the links between biological and cultural diversity carried out by UNESCO:

> *Just as cultural diversity needs to become an integral part of multilateral environmental agreements, biological diversity needs to be taken into consideration in political instruments dealing with culture and cultural diversity. A mechanism to link the separately evolving diversity agendas needs to be developed.'* (UNESCO, 2008, p34)

These and other germane statements are significant signs of progress in the recognition of the importance of culture and cultural diversity, along with biodiversity, for sustainability, and point to key implications for policy and implementation. In actual fact, however, practice is lagging behind: we are still far from a fully fledged acknowledgement of the concrete implications of the cultural diversity/biodiversity nexus for biodiversity conservation and for sustainability in its broadest sense. In many international processes and other relevant policy-making contexts (including the IUCN 2009–2012 Programme of Work), the model in use is by and large the one established and made popular by the Millennium Ecosystem Assessment (2005). In this assessment's framework, ecosystem services are the central concept, and the interdependence of ecological and socio-cultural dimensions is not explicitly recognized. Culture only comes into limited play under the rubric of 'cultural services' (aesthetic, spiritual, educational, recreational) – that is, as one category of 'services' that ecosystems provide for human well-being.

While there is no doubt that ecosystem health underpins human health and well-being, including aesthetic, spiritual and other cultural values, the reverse is also true (Rapport and Maffi, 2010). The ecosystem services framework does not account for the latter, inverse relationship: it reflects the ways in which the state of biodiversity and ecosystems affects people (including cultural values), but it does not reflect the ways in which the state of people (including cultural well-being) affects biodiversity and ecosystems. In other words, the framework does not include a feedback loop from cultural values, beliefs, knowledge, practices and languages to biodiversity – a loop that constitutes the very essence of the interdependence of cultural and biological diversity.

Closing this loop, so that this biocultural interdependence can be fully taken into account – that is, not only in theory but also in practice, with all the implications this has for policy and implementation – is the next, certainly momentous, step to be taken. What still stands in the way of completing the mind shift needed to close the loop is, in part, the sort of ideological and professional obstacles we described in Chapter 2. To these, of course, must be added major political and economic roadblocks – well encapsulated by Adams (2006, pp14–15): 'the immediate short-term interests of non-destitute citizens, businesses locked into current markets, financial institutions that believe they have no role beyond maintaining shareholder value, and timid politicians'.

These challenges notwithstanding, if we agree that 'sustainability is the path that allows humanity as a whole to maintain and extend quality of life through diversity of life' (Adams, 2006, p13), then to achieve sustainability we need to explicitly incorporate an expanded understanding of 'diversity of life' in this definition: an understanding of diversity as 'diversity of life in nature and culture'. This expanded understanding is what the biocultural perspective embodies. We need to take all actions required – at political, economic, social and institutional levels – to remove the obstacles that hamper our progress along the path toward that goal.

As Adams (2006, p15) notes, 'to have credibility and success, environmentalists need to move beyond the comfort zone of their established professional rituals and partnerships', because 'the changes needed cannot be brought about by environmentalists alone'. This calls for a 'rejuvenation' of the global environmental movement to commit 'to a path of justice and global equity' – the latter seen as central to any transition to sustainability (Adams and Jeanrenaud, 2008, p4). New partnerships will necessarily have to include genuine, rights-based, equal and equitable collaborations between conservationists and indigenous and local communities – collaborations that will fully recognize the interdependence of biological and cultural diversity, and deploy all means necessary to support both.

Figure 8.1 *Chake Chake village, Tanzania. In the perpetuation of biocultural diversity lies hope for the future of sustainability*

Source: Samantha Ross

The examples of integrated biocultural conservation projects presented in this volume – along with the many other similar initiatives underway worldwide, including Indigenous and Community Conserved Areas – should provide inspiration for the establishment of a new 'comfort zone' in the work of environmental conservation: comfort with the idea of the 'inextricable link' between the biological and the cultural diversity of life, and with what that implies for how conservation is done. We also hope that the lessons and recommendations we have drawn from these projects will help foster the development of policies and action plans by international and national agencies that will fully embrace and support the integrated protection, maintenance and restoration of diversity in both nature and culture. The future of sustainability of all life on Earth requires no less.

References

Abbi, A. (2006) 'Endangered Languages of the Andaman Islands', *Lincom Studies in Asian Linguistics*, vol 64, Lincom Europa, Munich, Germany

Abbi, A., Som, B. and Das, A. (2007) 'Where Have All the Speakers Gone? A sociolinguistic study of the Great Andamanese', *Indian Linguistics*, vol 68, no 3–4, pp325–343

Adams, W. M. (2006) 'The Future of Sustainability: Re-thinking environment and development for the twenty-first century', *Report of the IUCN Renowned Thinkers Meeting*, 29–31 January (2006), IUCN, Gland, Switzerland

Adams, W. M. and Jeanrenaud, S. J. (2008) *Transition to Sustainability: Towards a Humane and Diverse World*, IUCN, Gland, Switzerland

Alcorn, J. B. (1993) 'Indigenous Peoples and Conservation', *Conservation Biology*, vol 7, no 2, pp424–426

Alcorn, J. B. (1996) 'Is Biodiversity Conserved by Indigenous Peoples?', in S. K. Jain (ed.), *Ethnobiology in Human Welfare*, pp234–238, Deep Publications, New Delhi

Alcorn, J. B. and Royo, R. G. (2007) 'Conservation's Engagement with Human Rights: "Traction", "slippage", or avoidance?', *Policy Matters*, vol 15, pp115–139

Atran, S. (2001) 'The Vanishing Landscape of the Petén Maya Lowlands People: Plants, animals, places, words, and spirits', in Maffi (2001), pp157–174

Atran, S. and Medin, D. (1997) 'Knowledge and Action: Cultural models of nature and resource management in Mesoamerica', in M. Bazerman, D. Messick, A. Tinbrunsel and K. Wayde-Benzoni (eds), *Environment, Ethics, and Behavior*, pp171–208, New Lexington Press, San Francisco, CA

Aumeeruddy, Y. (1994) 'Local Representations and Management of Agroforests on the Periphery of Kerinci Seblat National Park, Sumatra, Indonesia', People and Plants Working Paper no 3

Aurukun Ethnobiology Database Project (2006) *Environment, Development and Sustainability*, vol 8, no 4, Springer, Netherlands, www.springerlink.com/content/q333704757u11756/fulltext.pdf

Bahn, P. and Flenley, J. R. (1992) *Easter Island, Earth Island*, Thames and Hudson, London

Balée, W. (1993) 'Indigenous Transformation of Amazonian Forests', *L'Homme*, vol 126–128, pp231–254

Balée, W. (1994) *Footprints of the Forest: Ka'apor Ethnobotany – The Historical Ecology of Plant Utilization by an Amazonian People*, Columbia University Press, New York

Baranyi, S. and Weitzner, V. (2006) 'Transforming Land-Related Conflict: Policy, practice and possibilities: policy brief', North–South Institute, Ottawa, ON

Barthlott, W., Biedinger, N., Braun, G., Feig, F., Kier, G. and Mutke, J. (1999) 'Terminological and Methodological Aspects of the Mapping and Analysis of Global Biodiversity', *Acta Botanica Finnica*, vol 162, pp103–110

Basso, K. (1996) *Wisdom Sits in Places*, University of New Mexico Press, Albuquerque, NM

Beltran, J. (ed.) (2000) *Indigenous and Traditional Peoples and Protected Areas: Principles, Guidelines, and Case Studies*, IUCN and WWF International, Gland, Switzerland

Berkes, F. (1999) *Sacred Ecology: Traditional Ecological Knowledge and Resource Management Systems*, Taylor and Francis, Philadelphia

Berkes, F. and Folke, C. (1994) 'Investing in Cultural Capital for Sustainable Use of Natural Resources', in S. Koskoff (ed.), *Investing in Natural Capital: The Ecological Economics Approach to Sustainability*, pp128–149, Island Press, Washington, DC

Berkes, F. and Folke, C. (eds) (1998) *Linking Social and Ecological Systems: Management Practices and Social Mechanisms for Building Resilience*, Cambridge University Press, Cambridge, UK

Berlin, B. (1992) *Ethnobiological Classification: Principles of Categorization of Plants and Animals in Traditional Societies*, Princeton University Press, Princeton, NJ

Bernard, P. S. (2003) 'Ecological Implications of Water Spirit Beliefs in Southern Africa: The need to protect knowledge, nature, and resource rights', USDA Forest Service, Proceedings RMRS-P-27

Bernard, R. (1992) 'Preserving Language Diversity', *Human Organization*, vol 51, no 1, pp82–89

Blythe, J, and McKenna Brown, R. (eds) (2004) 'Maintaining the Links. Language, identity and the Land', *Proceedings of the 7th Conference of the Foundation for Endangered Languages, Broome, Australia*, Foundation for Endangered Languages, Bath, UK

Bodley, J. H. (1990) *Victims of Progress*, Mayfield Publishing Co., Mountain View, CA

Borrini-Feyerabend, G., Pimbert, M., Farvar, M. T., Kothari, A. and Renard, Y. (2004) *Sharing Power: Learning-by-Doing in Co-management of Natural Resources Throughout the World*, IIED, IUCN/CEESP/CMWG, Cenesta, Tehran, Iran

Boster, J. (1984) 'Classification, Cultivation, and Selection of Aguaruna Cultivars of *Manihot esculenta* (Euphorbiaceae)', in G. Prance and J. Kallunki (eds), *Ethnobotany in the Neotropics*, pp34–47, Advances in Economic Botany, vol 1, New York Botanical Garden Press, Bronx, NY

Brosius, J. P. (1997) 'Endangered Forest, Endangered People: Environmentalist representations of indigenous knowledge', *Human Ecology*, vol 25, no 1, pp47–69

Brown, J., Mitchell, N. and Beresford, M. (eds) (2005) *The Protected Landscape Approach: Linking Nature, Culture, and Community*, IUCN, Gland, Switzerland and Cambridge, UK

Brush, S. B. (1980) 'Potato Taxonomies in Andean Agriculture', in D. Brokensha, D. M. Warren, and O. Werner (eds), *Indigenous Knowledge Systems and Development*, pp37–48, University Press of America, Lanham

Burney, D. A. and Flannery, T. F. (2005) 'Fifty Millennia of Catastrophic Extinctions after Human Contact', *Trends in Ecology and Evolution*, vol 20, no 7, pp395–401

Callicott, J. B. (1994) *Earth's Insights: A Survey of Ecological Ethics from the Mediterranean Basin to the Australian Outback*, University of California Press, Berkeley, CA

Campbell B. M. and Shackleton S. (2001) 'The Organizational Structures for Community-based Natural Resource Management in Southern Africa', *African Studies Quarterly*, vol 5, www.africa.ufl.edu/asq/v5/v5i3a6.htm

Carlson, T. J. S. and Maffi, L. (eds) (2004) *Ethnobotany and Conservation of Biocultural Diversity*, Advances in Economic Botany, vol 15, New York Botanical Garden Press, Bronx, NY

Chandler, M. J. and Lalonde, C. E. (1998) 'Cultural Continuity as a Hedge against Suicide in Canada's First Nations', *Transcultural Psychiatry*, vol 35, no 2, pp193–211

Chapin, M. (1992) 'The Co-existence of Indigenous Peoples and Environments in Central America', map insert, *Research and Exploration*, vol 8, no 2, Revised and reprinted as 'Indigenous Peoples and Natural Ecosystems in Central America and Southern Mexico', map insert, *National Geographic Magazine*, February 2003

Chapin, M. (2004) 'A Challenge to Conservationists', *World Watch* Nov/Dec, pp17–31

Chicchón, A. (2000) 'Conservation Theory Meets Practice', *Conservation Biology*, vol 14, no 5, pp1368–1369

Cochrane, P. (2006) 'Exploring Cultural Capital and its Importance in Sustainable Development', *Ecological Economics*, vol 57, pp318–330

Cocks, M. L. (2006) *Wild Resources and Practices in Rural and Urban Households in South Africa: Implications for Bio-cultural Diversity Conservation.* PhD Thesis, Wageningin University, The Netherlands, http://library.wur.nl/wda/dissertations/dis4026.pdf

Colchester, M. (2004) 'Conservation Policy and Indigenous Peoples', *Environmental Science and Policy*, vol 7, pp145–153

Colchester, M., Griffiths, T., Mackay, F. and Nelson, J. (2004) 'Indigenous Land Tenure: Challenges and possibilities', *Land Reform: Land Settlement and Cooperatives*, FAO, Rome

Collard, I. F. and Foley, R. A. (2002) 'Latitudinal Patterns and Environmental Determinants of Recent Human Cultural Diversity: Do humans follow biogeographical rules?', *Evolutionary Ecology Research*, vol 4, pp 371–383

Condit, R., Robinson, W. D., Ibánez, R., Aguilar, S., Sanjur, A., Martínez, R., Stallard, R. García, T., Angehr, T., Petit, L., Wright, S. J., Robinson, T. R. and Heckadon, S. (2001) 'Maintaining the Canal while Conserving Biodiversity around it: A challenge for economic development in Panama in the 21st century', *Bioscience*, vol 51, pp389–398

Conklin, B. and Graham, L. (1995) 'The Shifting Middle Ground: Amazonian Indians and eco-politics', *American Anthropologist*, vol 97, no 4, pp695–710

Crutzen, P. J. and Stoermer, E. F. (2000) 'The "Anthropocene"', *Global Change Newsletter*, vol 4, pp12–13

Dasmann, R. F. (1991) 'The Importance of Cultural and Biological Diversity', in M. L. Oldfield and J. B. Alcorn (eds), *Biodiversity: Culture, Conservation, and Ecodevelopment*, pp7–15, Westview Press, Boulder, CO

Davis, W. (2001) *Light at the Edge of the World*, Douglas and McIntyre, Vancouver, BC

Denevan, W. M. and Padoch, C. (eds) (1987) *Swidden-Fallow Agroforestry in the Peruvian Amazon*, Advances in Economic Botany, vol 5, New York Botanical Garden Press, Bronx, NY

Diamond, J. M. (1991) *The Rise and Fall of the Third Chimpanzee*, Harper and Collins, New York, NY

Diamond, J. M. (1993) 'Speaking with a Single Tongue', *Discover*, February, pp78–85

Diamond, J. M. (2005) *Collapse: How Societies Choose to Fail or Succeed*, Penguin Books, London

Dowie, M. (2005) 'Conservation Refugees', *Orion Magazine*, Nov/Dec

Eldredge, N. (1995) *Dominion*, H. Holt, New York, NY

Ellen, R. (1994) 'Rhetoric, Practice and Incentive in the Face of Changing Times: A study in Nuaulu attitudes to conservation and deforestation', in K. Milton (ed.), *Environmentalism: The View from Anthropology*, pp127–143, Routledge, London and New York, NY

Ellen, R. (2006) *Ethnobiology and the Science of Humankind*, Blackwell Publishing, Oxford

Ellis, D. W. (2001) 'Assessment of Recently Prepared Status Reports of the Committee on the Status of Endangered Wildlife in Canada for the inclusion of Aboriginal Traditional Knowledge, Community Knowledge, and Historical Data', Unpublished report for Environment Canada, Canadian Wildlife Service, Whitehorse, YK

Ellis, E. C. and Ramankutty, N. (2008) 'Putting People in the Map: Anthropogenic biomes of the world', *Frontiers in Ecology and the Environment*, vol 6

Ericksen, P. and Woodley, E. (2005) 'Using Multiple Knowledge Systems: Benefits and challenges', *Millennium Ecosystem Assessment*, Multiscale Assessments, vol 4, pp85–117, Island Press, Washington, DC

Esmail, T. (1997) 'Designing and Scaling-up Productive Natural Resource Management Programs: Decentralization and institutions for collective action', Decentralization, fiscal systems and rural development program, *Technical Consultation on Decentralization*, 16–18 December 1997, FAO, Rome

Fabricius, C. and Koch, E. (2004) 'Introduction', in C. Fabricus and E. Koch with H. Magome and S. Turner (eds), *Rights, Resources and Rural Development: Community-Based Natural Resource Management in Southern Africa*, ppxiii–xv, Earthscan, London and Sterling

Fishman, J. A. (1982) 'Whorfianism of the Third Kind: Ethnolinguistic diversity as a worldwide societal asset', *Language in Society*, vol 11, pp1–14

Flannery, T. (1995) *The Future Eaters: An Ecological History of the Australasian Lands and People*, George Braziller, New York, NY

Florey, M. (2006) 'Assessing the Vitality of Endangered Languages in Central Maluku', in Linguistic Society of the Philippines and ICAL (eds), *Tenth International Conference on Austronesian Linguistics (10th ICAL)*, Puerto Princesa City, Palawan, the Philippines

Fundación Hemera (2006) 'Concepto Etnológico sobre la etnicidad de las comunidades de Cota, Chía y Sesquilé que se reivindican Muiscas', Informe Final Ejecutivo (Contrato 342), Corporación Autónoma Regional de Cundinamarca, pp37, April 2006, Bogotá, Colombia

Gadgil, M., Subhash Chandran, M. D., Pramod, P., Ghate, U., Iyer, P., Gokhale, Y., Thomas, D. W. and Menon, P. (2002) 'People's Biodiversity Register, a Record of India's Wealth', Paper contributed to the workshop *Participatory Assessment, Monitoring and Evaluation of Biodiversity*, convened by the Environmental Change Institute, University of Oxford and held by Internet 7– 25 January 2002, available at www.etfrn.org/etfrn/workshop/biodiversity/index.html

Gragson, T. L. and Blount, B. G. (eds) (1999) *Ethnoecology: Knowledge, Resources, and Rights*, University of Georgia Press, Athens, GA and London

Grenville Goodwin Placenames Project (1997) 'Unpublished report to the Western Apache Coalition', White Mountain Apache Tribal Heritage Program, Whiteriver, AZ

Grimes, B. F. (1992) *Ethnologue: Languages of the World*, 12th edn, SIL International, Dallas, TX

Grin, F. (2005) 'The Economics of Language Policy Implementation: Identifying and measuring costs', in N. Alexander (ed.), *Proceedings of the Symposium 'Mother-Tongue Based Education in Southern Africa: The Dynamics of Implementation'*, pp11–25, PRAESA/Multilingualism Network, Cape Town

Groombridge, B. (ed.) (1992) *Global Biodiversity: Status of the Earth's Living Resources*, Chapman & Hall, London

Hames, R. B. (1991) 'Wildlife Conservation in Tribal Societies', in M. L. Oldfield and J. B. Alcorn (eds), *Biodiversity: Culture, Conservation, and Ecodevelopment*, pp172–199, Westview Press, Boulder, CO

Hardison, P. (2005) 'Commentary: Traditional knowledge studies and the indigenous trust', *Practicing Anthropology*, vol 27, no 1, pp42–45

Harmon, D. (1996) 'Losing Species, Losing Languages: Connections between biological and linguistic diversity', *Southwest Journal of Linguistics*, vol 15, pp89–108

Harmon, D. (1998) 'The Other Extinction Crisis: Declining cultural diversity and its implications for protected area management', in N. W. P. Munro and J. H. M. Willison (eds), *Linking Protected Areas with Working Landscapes Conserving Biodiversity*, Proceedings of the Third International Conference on Science and the Management of Protected Areas (SAMPA III), Calgary, AB, Canada, 12–16 May 1997, pp352–359, Science and Management of Protected Areas Association, Wolfville, NS

Harmon, D. (2002) *In Light of Our Differences: How Diversity in Nature and Culture Makes Us Human*, Smithsonian Institution Press, Washington, DC

Harmon, D. (2007) 'A Bridge over the Chasm: Finding ways to achieve integrated natural and cultural heritage conservation', *International Journal of Heritage Studies*, vol 13, no 4/5, pp380–392

Harmon, D. and Loh, J. (2004) 'The IBCD: A measure of the world's biocultural diversity', *Policy Matters*, vol 13, pp271–280

Harmon, D. and Loh, J. (2009) 'The Index of Linguistic Diversity', Technical report to The Christensen Fund, Terralingua

Heckenberger, M., Kuikuro, A., Kuikuro, U. T., Russell, J. C., Schmidt, M, Fausto, C. and Franchetto, B. (2003) 'Amazonia 1492: Pristine forest or cultural parkland?' *Science*, vol 30, pp1710–1714

Heckenberger, M. J., Russell, J. C., Toney, J. R. and Schmidt, M. J. (2007) 'The Legacy of Cultural Landscapes in the Brazilian Amazon: Implications for biodiversity', *Philosophical Transactions of the Royal Society B*, vol 362, pp197–208

Heywood, V. H. (ed.) (1995) *Global Biodiversity Assessment*, Cambridge University Press, Cambridge and New York

Human Rights Council (2006) 'Resolution 2006/2', UN Draft Declaration on the Rights of Indigenous Peoples

Hunn, E. S. (1999) 'The Value of Subsistence for the Future of the World', in V. Nazarea (ed.), *Ethnoecology: Situated Knowledge/Located Lives*, pp2–36, University of Arizona Press, Tucson, AZ

Hyndman, D. (1994) 'Conservation through Self-determination: Promoting the interdependence of cultural and biological diversity', *Human Organization*, vol 53, no 3, pp96–302

ICSU (2002) *Science and Traditional Knowledge*, Report from the ICSU Study Group on Science and Traditional Knowledge, March 2002, available at www.icsu.org/Gestion/img/ICSU_DOC_DOWNLOAD/220_DD_FILE_Traitional_Knowledge_report.pdf

IIFB (2006) 'Cross-cutting Approaches to the Implementation and Monitoring of the Goals of MDG 7: Indicators Relevant for Indigenous Peoples and the Convention on Biological Diversity', Permanent Forum on Indigenous Issues Fifth Session: United Nations, New York, NY

IUCN (2008) *Shaping a Sustainable Future: The IUCN Programme 2009–2012*, IUCN, Gland, Switzerland

Johnson, A. (1989) 'How the Machiguenga Manage Resources: Conservation or exploitation of nature?', in Posey and Balée (1989), pp213–222

Kassam, K.-A. S. (2009) *Biocultural Diversity and Indigenous Ways of Knowing: Human Ecology in the Arctic*, University of Calgary Press, Calgary, AB

Kellert, S. R., Mehta, J. N., Ebbin, S. A. and Lichtenfeld, L. L. (2000) 'Community Natural Resource Management: Promise, rhetoric, and reality', *Society and Natural Resources*, vol 13, pp705–715

Kenrick, J. and Lewis, J. (2004) 'Indigenous peoples' rights and the politics of the term "indigenous"', *Anthropology Today*, vol 20, no 2, pp4–9

Kepe T. (1999) 'The Problem of Defining "Community": Challenges for land reform programmes in rural South Africa', *Development Southern Africa*, vol 16, no 3, pp415–434

Kirch, P. V. (1997) 'Microcosmic Histories: Island perspectives on "global" change', *American Anthropologist*, vol 99, no 1, pp30–42

Kirch, P. V. and Hunt, T. L. (eds) (1996) *Historical Ecology in the Pacific Islands: Prehistoric Environmental and Landscape Change*, Yale University Press, New Haven, CT

Krauss, M. (1992) 'The World's Languages in Crisis', *Language*, vol 68, pp4–10

Krauss, M. (1996) 'Linguistics and Biology: Threatened linguistic and biological diversity compared', in *CLS 32, Papers from the Parasession on Theory and Data in Linguistics*, pp69–75, Chicago Linguistic Society, Chicago, IL

Krupnik, I. and Jolly, D. (eds) (2002) *The Earth Is Faster Now: Indigenous Observations of Arctic Environmental Change*, Arctic Research Consortium of the United States, Fairbanks, Alaska

Kuper, A. (2003) 'The Return of the Native', *Current Anthropology*, vol 44, no 3, pp396–397

Lizarralde, M. (2001) 'Biodiversity and Loss of Indigenous Languages and Knowledge in South America', in Maffi 2001, pp265–281

Loh, J. and Harmon, D. (2005) 'A Global Index of Biocultural Diversity', *Ecological Indicators*, vol 5, pp231–241

Long, J., Tecle, A. and Burnette, B. (2003) 'Cultural Foundations for Ecological Restoration on the White Mountain Apache Reservation', *Ecology and Society*, vol 8, no 1, available at www.consecol.org/vol8/iss1/art4/

López-Zent, E. and Zent, S. (2004) 'Amazonian Indians as Ecological Disturbance Agents: The Hoti of the Sierra de Maigualida, Venezuelan Guayana', in Carlson and Maffi (2004), pp79–112

Mace, R. and Pagel, M. (1995) 'A Latitudinal Gradient in the Density of Human Languages in North America', *Proceedings of the Royal Society of London B: Biological Sciences*, vol 261, pp117–121

Macintyre, M. and Foale, S. (2004) 'Politicized Ecology: Local responses to mining in Papua New Guinea', *Oceania*, vol 74

Maffi, L. (1998) 'Language: A resource for nature', *Nature and Resources: The UNESCO Journal on the Environment and Natural Resources Research*, vol 34, no 4, pp12–21

Maffi, L. (ed.) (2001) *On Biocultural Diversity: Linking Language, Knowledge, and the Environment*, Smithsonian Institution Press, Washington, DC

Maffi, L. (2005) 'Linguistic, Cultural, and Biological Diversity', *Annual Review of Anthropology*, vol 29, pp599–617

Maffi, L. (2007a) 'Biocultural Diversity and Sustainability', in J. Pretty, A. Ball, T. Benton, J. Guivant, D. Lee, D. Orr, M. Pfeffer and H. Ward (eds), *Sage Handbook on Environment and Society*, pp267–277, Sage Publications, Los Angeles, London, New Delhi and Singapore

Maffi, L. (2007b) 'Biocultural Diversity for Endogenous Development: Lessons from research, policy, and on-the-ground experiences', in B. Haverkort and S. Rist (eds), *Endogenous Development and Bio-Cultural Diversity: The Interplay of Worldviews, Globalisation and Locality*, pp56–66, COMPAS, Leusden, The Netherlands

Maffi, L. and Woodley, E. (2007) 'Culture', in Chapter 5, 'Biodiversity', *Global Environment Outlook: Environment for Development* (GEO 4), pp182–185, UNEP, Nairobi

Maffi, L., Skutnabb-Kangas, T. and Andrianarivo, J. (1999) 'Linguistic Diversity', in Posey (1999), pp21–57

Manne, L. L. (2003) ' Nothing Has Yet Lasted Forever: Current and threatened levels of biological diversity', *Evolutionary Ecology Research*, vol 5, pp517–527

Marsh, D. M., Trenham, P. C. and Colchester, M. (2000) 'Self-determination or Environmental Determinism for Indigenous Peoples in Tropical Forest Conservation', *Conservation Biology*, vol 14, no 5, pp1365–1367

Marton-Lefèvre, J. (2008) *Report of the Director General on the Work of the Union since the IUCN World Conservation Congress*, Bangkok (2004), IUCN, Gland, Switzerland

McNeely, J. A. and Keeton, W. S. (1995) 'The Interaction between Biological and Cultural Diversity', in B. von Droste, H. Plachter and M. Rossler (eds), *Cultural Landscapes of Universal Value: Components of a Global Strategy*, pp25–37, Fischer Verlag and UNESCO, Jena and New York, NY

Medin, D. L. and Atran, S. (eds) (1999) *Folkbiology*, MIT Press, Cambridge, MA and London

Meilleur, B. (1994) 'In Search of "Keystone Societies"', in N. L. Etkin (ed.), *Eating on the Wild Side: The Pharmacologic, Ecologic, and Social Implications of Using Noncultigens*, pp259–279, University of Arizona Press, Tucson, AZ

Meya, W. (2006) Letter to *The Financial Times*, London, 11 March 2006

Millennium Ecosystem Assessment (2005) *Ecosystems and Human Well-Being: Synthesis Report*, Island Press, Washington, DC

Minnis, P. E. and Elisens, W. J. (eds) (2000) *Biodiversity and Native America*, University of Oklahoma Press, Norman, OK

Mittermeier, R. A., Myers, N., Thomsen, J. B., da Fonseca, G. A. B. and Olivieri, S. (1998) 'Biodiversity Hotspots and Major Tropical Wilderness Areas: Approaches to setting conservation priorities', *Conservation Biology*, vol 12, no 3, pp516–520

Moock, J. and Rhoades, R. (1992) *Diversity, Farmer Knowledge and Sustainability*, Cornell University Press, Ithaca, NY

Moore, J. L., Manne, L., Brooks, T., Burgess, N. D., Davies, R. et al (2002) 'The Distribution of Cultural and Biological Diversity in Africa', *Proceedings of the Royal Society of London B: Biological Sciences*, vol 269, pp1645–1653

Mühlhäusler, P. (1995) 'The Interdependence of Linguistic and Biological Diversity', in D. Myers (ed.), *The Politics of Multiculturalism in the Asia/Pacific*, pp154–161, Northern Territory University Press, Darwin, Australia

Nabhan, G. P. (1997) *Cultures of Habitat*, Counterpoint, Washington, DC

Nabhan, G. P., Joe, T., Maffi, L., Pynes, P., Seibert, D., Sisk, T. D., Stevens, L. E. and Trimble, S. (2002) *Safeguarding the Uniqueness of the Colorado Plateau: An Ecoregional Assessment of Biocultural Diversity*, Center for Sustainable Environments, Terralingua, and Grand Canyon Wildlands Council, Flagstaff, AZ

National Research Council (1989) *Lost Crops of the Incas: Little-Known Plants of the Andes with Promise for Worldwide Cultivation*, National Research Council, Washington, DC

Nations, J. D. (2001) 'Indigenous Peoples and Conservation: Misguided myths in the Maya tropical forest', in L. Maffi (ed.), *On Biocultural Diversity: Linking Language, Knowledge, and the Environment*, Smithsonian Institute, Washington, DC

Nettle, D. (1996) 'Language Diversity in West Africa: An ecological approach', *Journal of Anthropological Archaeology*, vol 15, pp403–438

Nettle, D. (1998) 'Explaining Global Patterns of Linguistic Diversity', *Journal of Anthropological Archaeology*, vol 17, pp354–374

Nettle, D. (1999) *Linguistic Diversity*, Oxford University Press, Oxford

Nichols, J. (1990) 'Linguistic Diversity and the First Settlement of the New World', *Language*, vol 66, pp475–521

Nichols, J. (1992) *Linguistic Diversity in Space and Time*, University of Chicago Press, Chicago/London

Nietschmann, B. Q. (1992) *The Interdependence of Biological and Cultural Diversity*, Occasional Paper, no 21, Center for World Indigenous Studies, December

Oldfield, M. L. and Alcorn, J. (1987) 'Conservation of Traditional Agroecosystems', *BioScience*, vol 37, no 3, pp199–208

Olson, D. M. and Dinerstein, E. (1998) 'The Global 200: A representation approach to conserving the Earth's most biologically valuable ecoregions', *Conservation Biology*, vol 12, no 3, pp502–515

Oviedo, G., Maffi, L. and Larsen, P. B. (2000) *Indigenous and Traditional Peoples of the World and Ecoregion Conservation: An Integrated Approach to Conserving the World's Biological and Cultural Diversity*, WWF-International and Terralingua, Gland, Switzerland

Pannell, S. (2006) 'Reconciling Nature and Culture in a Global Context? Lessons from the World Heritage List', Report No. 48, available at www.rainforest-crc.jcu.edu.au/publications/nature_culture.htm

Pattanayak, D. P. (1988) 'Monolingual Myopia and the Petals of the Indian Lotus: Do many languages divide or unite a nation?' in T. Skutnabb-Kangas and J. Cummins (eds), *Minority Education: From Shame to Struggle*, pp379–389, Multilingual Matters, Clevedon, UK

Pikangikum First Nation and Ontario Ministry of Natural Resources (2006) 'Keeping the Land: A land use strategy for Whitefeather Forest and adjacent areas', available at www.whitefeatherforest.com/pdfs/land-use- strategy.pdf

Ploeg, J. van der and Weerd, M. van (2004) 'Devolution of Natural Resource Management and Philippine Crocodile Conservation in San Mariano, Isabela', *Philippine Studies*, vol 52, no 3, pp346–383

Ploeg, J. van der, Cureg, M. C. and Weerd, M. van (2008) 'Mobilizing Public Support for In-situ Conservation of the Philippine Crocodile in the Northern Sierra Madre: Something to be proud of!', *National Museum Papers*, vol 14, pp68–94

Poland M., Hammond-Tooke, D. and Voigt, L. (2003) *The Abundant Herds: A Celebration of the Nguni Cattle of the Zulu people*, Fernwood Press, Vlaeberg

Ponting, C. (1991) *A Green History of the World*, Sinclair-Stevenson, London

Posey, D. A. (1984) 'A Preliminary Report on Diversified Management of Tropical Forest by Kayapó Indians of the Brazilian Amazon', in G.T. Prance and J. A. Kallunki (eds), *Ethnobotany in the Neotropics*, pp112–126, Advances in Economic Botany, vol 1, New York Botanical Garden Press, Bronx, NY

Posey, D. (1998) 'Diachronic Ecotones and Anthropogenic Landscapes in Amazonia: Contesting the consciousness of conservation', in W. Balée (ed.), *Advances in Historical Ecology*, pp104–118, Columbia University Press, New York, NY

Posey, D. A. (ed.) (1999) *Cultural and Spiritual Values of Biodiversity*, Intermediate Technology Publications and UNEP, London and Nairobi

Posey, D. A. and Balée, W. (eds) (1989) *Resource Management in Amazonia: Indigenous and Folk Strategies*, Advances in Economic Botany, vol 7, New York Botanical Garden Press, Bronx, NY

PRATEC (1998) *La crianza ritual de las semillas en los Andes*, PRATEC, Lima, p4

PRATEC (2006) *Reflections on the In Situ Project II, Lima*, PRATEC, pp37–38

Rapport, D. J. (2007) 'Healthy Ecosystems: An evolving paradigm', in J. Pretty, A. Ball, T. Benton, J. Guivant, D. Lee, D. Orr, M. Pfeffer and H. Ward (eds), *Sage Handbook on Environment and Society*, pp431–441, Sage Publications, Los Angeles, London, New Delhi and Singapore

Rapport, D. J. (2008) 'How Are We Doing? Finding ways in which humans can live in harmony with Nature', Web exclusive, *Resurgence Magazine*, vol 250, September/October 2008

Rapport, D. J. and Maffi, L. (2010) 'The Dual Erosion of Biological and Cultural Diversity: Implications for the health of eco-cultural systems', to appear in J. Pretty and S. Pilgrim (eds), *Nature and Culture: Revitalizing the Connection*, Earthscan, London

Rapport, D. J. and Singh, A. (2006) 'An Ecohealth-based Framework for State of Environment Reporting', *Ecological Indicators*, vol 6, pp409–428

Redford, K. H. (1990) 'The Ecologically Noble Savage', *Orion*, vol 9, no 3, pp24–29

Redford, K. H. and Brosius, J. P. (2006) 'Diversity and Homogenization in the Endgame', *Global Environmental Change*, vol 16, pp317–319

Redford, K. H. and Stearman, A. M. (1993) 'Forest-dwelling Native Amazonians and the Conservation of Biodiversity: Interests in common or in collision?', *Conservation Biology*, vol 7, no 2, pp248–255

Redford, K. H. and Sanderson, S. E. (2000) 'Extracting Humans from Nature', *Conservation Biology*, vol 4, no 5, pp1362–1364

Rengifo, G. (2005) 'The Educational Culture of the Andean-Amazonian Community', in *Interculture*, no 148, April, Intercultural Institute of Montreal, Montréal, PQ, pp3–46

Riddell, J. C. (2000) 'Emerging Trends in Land Tenure Reform: Progress towards a unified theory', FAO Land Tenure Service, Sustainable Development Department FAO, Rome

Robinson, G. S., Pigott Burney, L. and Burney, D. A. (2005) 'Landscape Paleoecology and Megafaunal Extinction in Southeastern New York State', *Ecological Monographs*, vol 75, no 3, pp295–315, The Ecological Society of America

Romero, C. and Andrade, G. I. (2004) 'International Conservation Organizations and the Fate of Local Tropical Forest Conservation Initiatives', *Conservation Biology*, vol 18, no 2, pp578–580

Ross, N. (2002) 'Cognitive Aspects of Intergenerational Change: Mental models, cultural change, and environmental behavior among the Lacandon Maya of southern Mexico', *Human Organization*, vol 61, pp125–138

Schwartzman, S., Moreira, A. and Nepstad, D. (2000a) 'Rethinking Tropical Forest Conservation: Perils in parks', *Conservation Biology*, vol 14, no 5, pp1351–1357

Schwartzman, S., Nepstad, D. and Moreira, A. (2000b) 'Arguing Tropical Forest Conservation: People versus parks', *Conservation Biology*, vol 14, no 5, pp1370–1374

Selin, H. (ed.) (2003) *Nature Across Cultures: Views of Nature and the Environment in Non-Western Cultures*, Kluwer Academic Publishers, Dordrecht

Shepard, G. H., Jr. (2002) 'Primates in Matsigenka Subsistence and Worldview', in A. Fuentes and L. Wolfe (eds), *Primates Face to Face: The Conservation Implications of Human and Nonhuman Primate Interconnections*, pp101–136, Cambridge University Press, Cambridge, UK

Shepard, G. H., Jr. (2004) 'Ethnobotanical Ground-truthing and Forest Diversity in the Western Amazon', in Carlson and Maffi (2004), pp133–171

Skutnabb-Kangas, T., Maffi, L. and Harmon, D. (2003) *Sharing a World of Difference: The Earth's Linguistic, Cultural, and Biological Diversity* and map *The World's Biocultural Diversity: People, Languages, and Ecosystems*, UNESCO, Paris

Smith, E. A. (2001) 'On the Coevolution of Cultural, Linguistic, and Biological Diversity', in Maffi (2001), pp95–117

Soulé, M. E. (2000) 'Does Sustainable Development Help Nature?' *Wild Earth*, Winter, vol 2000–2001, pp57–64

Stavenhagen, R. (2000) 'Culture and Poverty', *WORLD 2000: Cultural Diversity, Conflict and Pluralism*, UNESCO, Paris

Stepp, J. R., Wyndham, F. S. and Zarger, R. (eds) (2002) *Ethnobiology and Biocultural Diversity*, University of Georgia Press, Athens, GA

Stepp, J. R., Cervone, S., Castaneda, H., Lasseter, A., Stocks, G. and Gichon, Y. (2004) 'Development of a GIS for Global Biocultural Diversity', *Policy Matters*, vol 13, pp267–270

Stepp, J. R., Castaneda, H. and Cervone, S. (2005) 'Mountains and Biocultural Diversity', *Mountain Research and Development*, vol 25, no 3, pp223–227

Stepp, J. R., Castaneda, H., Reilly-Brown, J. and Russell, C. (2008) *Set of 13 Maps of Global and Regional Biocultural Diversity*, Biocultural Diversity Mapping Project, University of Florida, Gainesville, FL

Stevens, S. (ed.) (1997) *Conservation Through Cultural Survival: Indigenous Peoples and Protected Areas*, Island Press, Washington, DC

Sutherland, W. J. (2003) 'Parallel Extinction Risk and Global Distribution of Languages and Species', *Nature*, vol 423, pp276–279

Tebtebba Foundation (2002) 'The Myth of Sustainable and Responsible Mining', *Indigenous Perspectives*, vol 1

Teilhard de Chardin, P. (1966) *Man's Place in Nature*, Collins, London [Translation of French original of 1956]

Terborgh, J. (1999) *Requiem for Nature*, Island Press, Washington, DC

Terborgh, J. (2000) 'The Fate of Tropical Forests: A matter of stewardship', *Conservation Biology*, vol 14, no 5, pp1358–1361

Thangaraj, K., Chaubey, G., Kivisild, T., Reddy, A. G., Singh, V. K., Rasalkar, A. A. and Singh, L. (2005) 'Reconstructing the Origin of Andaman Islanders', *Science*, vol 308, no 5724, pp996

Thrupp, L. A. (1998) *Cultivating Diversity*, World Resources Institute, Washington, DC

Toulmin, C. and Chambers, R. (1990) *Farmer-first: Achieving sustainable dryland development in Africa*, International Institute for Environment and Development, London

Tucker, C. M. (2004) 'Land Tenure Systems, and Indigenous Intellectual Property Rights', in M. Riley (ed.), *Indigenous Intellectual Property Rights*, pp127–151, AltaMira Press, Walnut Creek, CA

Tumanyire, G. (2002) 'Traditional and Modern Health Practitioners Together Against AIDS and other Diseases', *THETA News, Official Newsletter of Theta*, vol 7, no 1, Jan–June

UNDP (2004) *Human Development Report 2004: Cultural liberty in today's diverse world*, United Nations Development Programme, New York, NY

UNEP (2007) *Global Environment Outlook: Environment for Development*, UNEP, Nairobi

UNEP and UNESCO (2003) *Cultural Diversity and Biodiversity for Sustainable Development*, UNEP, Nairobi

UNESCO (2000) *World Conference on Science, Science for the Twenty-First Century: A New Commitment*, UNESCO, Paris

UNESCO (2003) 'Language Vitality and Endangerment', International Expert Meeting of the UNESCO Intangible Cultural Heritage Unit's Ad Hoc Expert Group on Endangered Languages, UNESCO, Paris-Fontenoy, 10–12 March, Document approved by the participants on 31 March 2003, available at www.unesco.org/culture/ich/doc/src/00120-ECN.pdf

UNESCO (2008) *Links Between Biological and Cultural Diversity: Concepts, Methods and Experiences*, Report of an International Workshop, UNESCO, Paris

UNESCO (2009) *Learning and Knowing in Indigenous Societies Today*, P. Bates, M. Chiba, S. Kube and D. Nakashima (eds), UNESCO, Paris

United Nations (2007) *Declaration on the Rights of Indigenous People*, United Nations General Assembly, Sixty-first Session, New York

United Nations University (2005) 'Establishing a UNU Initiative on Traditional Knowledge', United Nations University, *Institute for Advanced Studies*, vol 42

UNPFII (Permanent Forum on Indigenous Issues) (2006) *Report of the Meeting on Indigenous Peoples and Indicators of Well-Being, Ottawa, ON*, UNPFII, New York, NY

Valladolid, J. (1998) 'Andean Peasant Agriculture: Nurturing a diversity of life in the chacra' in Apffel Marglin, F. (ed.), *The Spirit of Regeneration: Andean Culture Confronting Western Notions of Development*, Zed Books, London

WCED (1987) *Our Common Future*, Report of the World Commission on Environment and Development, Oxford University Press, Oxford and New York, NY

Weber, R., Butler, J. and Larson, P. (eds) (2000) *Indigenous Peoples and Conservation Organizations: Experiences in Collaboration*, WWF-US, Washington, DC

Weerd, M. van and Ploeg, J. van der (2004) 'A New Future for the Philippine Crocodile', *Sylvatrop*, vol 13, nos 1 & 2, pp31–50

Wilcox, B. A. and Duin, K. N. (1995) 'Indigenous Cultural and Biological Diversity: Overlapping values of Latin American ecoregions', *Cultural Survival Quarterly*, vol 18, no 4, pp49–53

Williams, N. M. and Baines, G. (eds) (1993) *Traditional Ecological Knowledge: Wisdom for Sustainable Development*, Centre for Resource and Environmental Studies, National Australian University, Canberra

Wilson, M., Kaplan, S. and Maki, T. (1952) *Keiskammahoek Rural Survey Volume 3: Social Structure*, Shooter and Shooter, Pietermaritzburg

Woodley, E., Crowley, E., Dey de Pryck, J. and Carmen, A. (2008) *Cultural Indicators of Indigenous Peoples' Food and Agro-ecological Systems*, UN Food and Agricultural Organization (FAO) and the International Indian Treaty Council (IITC), FAO, Rome

World Rainforest Movement, The (1992) *Charter of the Indigenous-Tribal Peoples of the Tropical Forests*, Penang, Malaysia

Wright, R. (2004) *A Short History of Progress*, Anansi, Toronto, ON

Wroe, S., Field, J. and Grayson, D. K. (2006) 'Megafaunal Extinctions: Climate, humans and assumptions', *Trends in Ecology and Evolution*, vol 21, no 2, pp61–62

Wurm, S. (ed.) (2001) *Atlas of the World's Languages in Danger of Disappearing*, UNESCO, Paris

WWF, Zoological Society of London and Global Footprint Network (2008) *2010 and Beyond: Rising to the Biodiversity Challenge*, WWF, Gland, Switzerland

Yunnan Initiative, The (2000) 'The Yunnan Initiative: Visions and actions for the enhancement of biological and cultural diversity', Cultures and Biodiversity Congress, 20–30 July 2000, Kunming, Xishuangbanna, Deqing, PR China

Zarger, R. K. and Stepp, J. R. (2004) 'Persistence of Botanical Knowledge among Tzeltal Maya Children', *Current Anthropology*, vol 45, no 3, pp413–418

Zent, S. (1999) 'The Quandary of Conserving Ethnoecological Knowledge: A Piaroa example', in Gragson and Blount (1999), pp90–124

Zent, S. (2001) 'Acculturation and Ethnobotanical Knowledge Loss among the Piaroa of Venezuela: Demonstration of a quantitative method for the empirical study of TEK change', in Maffi (2001), pp190–211

Zent, S. (2008) 'Methodology for Developing a Vitality Index of Traditional Environmental Knowledge', Technical report to The Christensen Fund, Terralingua

Zent, S. and López-Zent, E. (2004) 'Ethnobotanical Convergence, Divergence, and Change among the Hoti of the Venezuelan Guayana', in Carlson and Maffi (2004), pp37–78

Zent, S. and López- Zent, E. (2007) 'On Biocultural Diversity from a Venezuelan Perspective: Tracing the interrelationships among biodiversity, culture change, and legal reforms', in C. McManis (ed.), *Biodiversity and the Law: Intellectual Property, Biotechnology and Traditional Knowledge*, pp91–114, Earthscan, London

Appendix 1
Analytical Tables

Ellen Woodley

Table A1.1 *Overview of the biodiversity and cultural conservation aspects of projects*

#	*Project name*	*Location*	*Biodiversity conservation*	*Cultural conservation*	*Language revitalization*	*Linkages made*
1	A Review of the Birds and Plants of Bikini Atoll, Trees of the Marshall Islands and Fish of Micronesia	Majuro, Marshall Islands	Names and uses of birds, trees and fish are recalled and documented years after evacuation for nuclear testing, to increase awareness of local biodiversity and provide incentive to conserve. Project also attempts to preserve the only remaining indigenous land bird in the Marshall Islands.	Preservation of traditional knowledge and classification of the natural environment. Transmission of local knowledge to younger generations strengthens cultural identity.	Local names of birds, molluscs, trees and fish species, showing the differences and similarities in the eight languages of the region, are published in booklets for schools.	Documentation of local knowledge (LK) and classification of biodiversity of Marshall Islands, where LK is seriously threatened, may assist in conservation when people return to the islands.
2	Indigenous Theory for Health: Enhancing Traditional-Based Indigenous Health Services in Vancouver	British Columbia, Canada	Ecological and socio-cultural resilience and sustainability are enhanced through working with indigenous healers and drawing on indigenous theories of holism, which apply to relations between people and environment to engender a greater appreciation of biodiversity. Project calls for protection of local and traditional medicine harvesting sites, sacred practice sites, and the development of environmental spaces for healing.	Recognition, support and validation of urban traditional indigenous healers/health services and their philosophy of engagement with others and the environment, based on holism.	Indigenous language revitalization is not the project's focus, however, the project operates on the principle that indigenous world views contain whole knowledge systems which are embedded in language as well as values, practices and material goods.	Project encourages the link between indigenous theories of holism and beneficial relationship to environment, which should contribute to sustainable ecosystem management, including biodiversity conservation.

#	*Project name*	*Location*	*Biodiversity conservation*	*Cultural conservation*	*Language revitalization*	*Linkages made*
3	Jande Myra Ta Ka'a Rupi Ha (Our Trees of the High Forest): Ka'apor Ethnodendrology	Eastern Amazonian Brazil	Defend borders of Alto Turiaçu Indigenous Reserve (Maranhão, Brazil) to guarantee its biodiversity along with its cultural diversity.	Dissemination and protection of the history, customs, artistic and traditional cultural practices of the Ka'apor people related to sustainable management of resources, culture and environmental protection.	Support for continuation of the Ka'apor language in its cultural context.	Drawing upon traditional practices to conserve and sustainably use local biodiversity on an indigenous reserve.
4	Bamenda Highlands Forest Project	Northwest Province, Cameroon	Project involves conserving the largest remaining montane forest in West Africa, home to a number of endemic species of animals and plants, most of which are highly threatened due to loss of habitat.	Project is based on local demand for forest conservation to maintain the remnants of forest for traditional uses. Protected community forests are designated and managed for traditional and spiritual values as well as utilitarian reasons.	Local languages are used for all species names; written materials are in local languages. While this was not an explicit objective, the frequent and ongoing discussions about the forest have helped to revive and pass on elements of language to more people, including younger generations.	Utilization of community initiatives based on traditional forest uses to conserve fragments of rare West African montane forest.
5	Biocultural Diversity: Elaborating Theoretical Issues for Communities and Policy Makers	Victoria and Northern Australia	Investigation of the feasibility of using local, Aboriginal indicators of biodiversity for database development. Sustainable resource management is dependent on local languages and cultures, which are being recognized and documented.	Seeks to devise forms of data assemblage that integrate practices, performance and ritual related to ecology. Recognizes the need to understand local and indigenous ecological knowledge in order to understand biodiversity and to have sustainable resource management. Strengthens cultural relationships to the environment.	Project aims to enhance the strength of local languages and to determine the role of narrative in knowledge assemblage and whether there are key material practices connecting narrative and land uses.	Project focuses on representing and strengthening traditional/indigenous worldviews for integrated resource management.

#	*Project name*	*Location*	*Biodiversity conservation*	*Cultural conservation*	*Language revitalization*	*Linkages made*
6	Eco-cultural Health in the Sierra Tarahumara, Mexico	Sierra Tarahumara, Chihuahua, Mexico	Assessment of the health of the local landscape and revegetation and restoration projects.	Culturally appropriate education for Rarámuri children and youth; improvement of water and food sources as well as health as basis for life and cultural survival.	Prevailing educational approach disfavours Rarámuri language and culture; project aims to determine how Rarámuri language might be integrated in alternative in-situ education initiatives.	Support to Rarámuri efforts to recover and take direct control over the eco-cultural health of their landscape and communities.
7	Collection and Documentation of Traditional Conservation Sites	Marshall Islands	Traditional beliefs related to conserved areas are being used to guide resource management and conserve local biodiversity today.	Traditional knowledge and beliefs (taboos) are used in management policy. Project examines relationship between traditional rulers and the traditional conservation system established on each island.	Many traditional Marshallese words being lost and/or replaced by foreign (English) words. Revitalizing traditional management practices will increase use of the associated language.	Use of traditional practices for conservation is integrated into resource management.
8	Conservation in Managed Indigenous Áreas (Conservación en Areas Indígenas Manejadas, CAIMAN)	Ecuador	Consolidating and enforcing the legal rights of indigenous peoples over their territory contributes to biodiversity conservation. Participatory management plans and economic alternatives are implemented to reduce the pressure on local fauna.	Strengthening of indigenous federations; legalizing of ancestral territories according to Ecuador's laws. Indigenous nationalities can gain title to their ancestral territories, and have a constitutional right to be consulted prior to the initiation of extractive activities within their territories. Conservation of traditional artisan skills through the promotion of handicrafts, whereby elders and artisans transmit cultural knowledge to others.	Not a project focus.	High level of participation ensures that cultural traditions and values become integrated into resource management plans that address biodiversity issues.

#	*Project name*	*Location*	*Biodiversity conservation*	*Cultural conservation*	*Language revitalization*	*Linkages made*
9	Crocodile Rehabilitation, Observance and Conservation	Northeast Luzon, Philippines	Project aims to conserve populations of the critically endangered Philippine crocodile in the wild.	Community-based conservation strategy for the Philippine crocodile, based on crocodiles' past persistence in the ancestral domains of the Kalinga and Agta people, where they were protected by a system of beliefs and taboos, which are now rapidly disappearing.	Not explicit, but all communication material for the awareness campaign is in the local languages Tagalog and Ilocano.	Support in obtaining land rights and efforts to strengthen and formalize the traditional ways of protecting the endangered crocodiles. Traditional practices that assisted Philippine crocodile conservation are revived and traditional knowledge on the behaviour and ecology of the crocodile is documented.
10	Dance for the Earth and for Her Peoples	Latin America, Caribbean, Africa and Europe	Dances to celebrate and promote renewed commitment for biodiversity conservation.	Many traditional dances have strong links to nature, landscape and conservation. They borrow movements from animals, express seasonal and annual cycles, or act out stories related to nature.	Project is not language oriented.	Performing arts are a tool for promoting the conservation of biocultural diversity at select protected areas around the globe. Dance is used to strengthen the links between conservation of nature and the maintenance of culture. Stories and dance celebrate the efforts of communities conserving their traditional lands.
11	Environmental Applications Reference Thesaurus (EARTh)	Italy	Thesaurus includes a conceptual and terminological segment specifically concerning biodiversity conservation.	Traditional knowledge classification systems and environmental terminologies encapsulate traditional worldviews and reflect indigenous cognitive structuring of reality, helping nourish the sense of cultural identity and better represent traditional cultures within the global context.	Developing a thesaurus documents specific language associated with biodiversity.	Using traditional environmental terminologies and classification systems to assist with classifying biodiversity for conservation purposes.

#	Project name	Location	Biodiversity conservation	Cultural conservation	Language revitalization	Linkages made
12	Establishing Marine Protected Areas and Spatio-temporal Refugia in the Roviana and Vona Vona Lagoons, Solomon Islands	Solomon Islands	Conserve marine and riparian habitats of flagship species, as well as sites where vulnerable or endangered marine species are found.	Project has created and consolidated Community-Based Marine Orotected Areas under customary land/sea tenure regimes (traditional authority and practices). Local knowledge is documented and utilized for conservation purposes.	Environmental dictionary in the Roviana vernacular describes all marine and terrestrial organisms known locally.	Traditional knowledge of endangered species and habitats as well as customary marine tenure systems are used to conserve species and ecosystems in a system of Community-Based Marine Protected Areas.
13	Forests and Oceans for the Future	British Columbia, Canada	Local sustainable forest and natural resource management strategies are applied in provincial management plans to sustainably use biodiversity.	Core community values and knowledge (indigenous classification and understandings of forests, marine plants and animals), as well as customary forms of governance among the Gitxaała that regulate human action within the environment, act to conserve and enhance biodiversity and lead to long-term sustainability within the Gitxaała traditional territory.	Designing curriculum materials and an indigenous field guide with local classifications.	Traditional knowledge and values as well as traditional governance systems contribute to resource management.
14	Gwich'in Place Names and Traditional Land Use	Northwest Territories and Yukon, Canada	Inventory of heritage sites and incorporation of extensive land use information into the Gwich'in Land Use Plan.	Gwich'in place names and the associated stories along with trails, traditional camp sites, graves, historic sites, harvesting locales and sacred or legendary places are windows into Gwich'in culture and history. Elders and youth are brought together on the land, to promote and pass on the language and knowledge about the land and the culture.	Recorded information on approximately 1000 named places, most of which are in the Gwich'in language. Project promotes passing language from elders to youth. Official recognition of Gwich'in place names on road signs and maps. There are language revitalization initiatives and a language immersion camp.	Integration of Gwich'in traditional knowledge of the land through the use of place names and associated stories to influence current sustainable land use plans.

#	*Project name*	*Location*	*Biodiversity conservation*	*Cultural conservation*	*Language revitalization*	*Linkages made*
15	Andean Project for Peasant Technologies (Proyecto Andino para las Tecnologías Campesinas, PRATEC)	Peru	Projects revive the concept of 'communities of nurturers of the diversity of plants and animals' in several biodiverse regions in Peru to conserve the diversity of native cultivated plants and their wild relatives in the central Andes. The role of children as nurturers of biodiversity is also explored.	Projects promote revitalization of traditional nurturing of plants/animals and local landscape via the agro-festive cycle: practices of soil preparation, sowing, harvesting, storage and food preparation are revived. The project found that agrobiodiversity is the result of Andean Amazonian agricultural practices, Therefore, strengthening local cosmovision and nurturing respect are fundamental to agrobiodiversity conservation.	Basis of school curriculum are the traditional agricultural practices of local communities in the local language. Community radio programmes are attempting to do more programming in Quechua.	Sophisticated, caring maintenance of agrobiodiversity carried out by the *campesino* communities includes respect for wider aspects of nature (including spiritual beliefs and values), which assists with conservation of native endangered species.
16	Jaru Ethnobiological Language Knowledge Project	Western Australia	Intensive fieldwork was done to document ethnobiological resources in the region.	Focus in on working with Jaru elders to document local ethnobiology and how language can be used to transmit information about the ecological landscape. Jaru elders' knowledge of trees used in artefact making and knowledge of bush medicines are being documented.	Kimberley Language Resource Centre is often asked to provide support to Kimberley language groups carrying out documentation of plants and animals.	Project was established to consolidate strong language transmission outcomes from ethnobiological documentation.

#	*Project name*	*Location*	*Biodiversity conservation*	*Cultural conservation*	*Language revitalization*	*Linkages made*
17	Linking Crop Diversity with Food Traditions and Food Security in the Hills of Nepal	Kaski, Nepal	Conservation of agricultural biodiversity is achieved by promoting traditions surrounding the diversity of crop landraces.	Project examines how traditional use of local crop varieties in traditional communication channels such as festivals and life-cycle rituals helps maintain use of traditional foods and agricultural biodiversity on the farm.	Information is generated in local language and translated into English. Project view is that revitalization of language in isolation is not a panacea for maintaining transmission of knowledge and practices.	Landraces are likely to be maintained on farm as long as the belief system and traditional practices continue in the social system.
18	Participatory Genetic Improvement of Traditional Crops and Native Tree Species	Costa Rica	Seed exchanges of endangered farm, forest and medicinal plant species as well as collection of diverse, locally adapted organic seeds to conserve genetic diversity.	Traditional practice of seed exchange and use of local knowledge associated with traditional crop seeds, endangered forest species and medicinal plants is documented with scientific verification.	Indigenous names of plants and traditional practices that store cultural information are documented.	Drawing on traditional practices of seed exchanges and knowledge of plant species promotes the use and maintenance of diverse genetic stock of several plant species.
19	Promotion of Traditional Medicine and Indigenous Cultural Research and African Spirituality	Kampala District, Uganda	Sustainable use and thus conservation of biodiversity by promoting traditional uses of plants and nurturing medicinal plant species.	Traditional beliefs and practices associated with medicinal plants are studied and documented among practitioners in a traditional healing demonstration institute. Promotes cultural pluralism.	Language conservation is not a focus, but the project encourages the documentation and recording of traditional information in local languages.	Cultural approach to biodiversity conservation through medicinal plant practitioners working together and sharing ideas and methods.
20	Support Project for the Ngäbe Indigenous People (Proyecto de Apoyo al Pueblo Indígena Ngäbe)	Costa Rica	Ngäbe reservations maintain about 70 per cent forest cover, consisting of a rich variety of habitats encompassing three of the five elevational zones found in Costa Rica. Premise is that resource management within the indigenous territory leads to biodiversity conservation.	Project assists the Ngäbe in reversing the loss of their culture, recovering traditional political institutions and traditional medicine, supporting the defence of territory and appropriate management practices and improving food production systems.	Ngäbe youth are learning to write in their original language; book on traditional medicinal plants written in Ngäbere language. Elders, indigenous teachers and Ministry of Education all contributed to establishing the written standard for the language.	Maintenance of traditional cultural practices is integrated with sustainable use of indigenous lands.

#	*Project name*	*Location*	*Biodiversity conservation*	*Cultural conservation*	*Language revitalization*	*Linkages made*
21	Social, Environmental and Economic Sustainability in the Context of Melanesian Mining Projects	New Ireland Province, Papua New Guinea	Environmental impact assessment associated with mining draws upon local and traditional ecological knowledge of local biodiversity.	Socio-cultural study of cultural loss from impacts of mining. Project also examines local understandings of environmental impact and use of local knowledge for the assessment of mining impacts, thus conserving aspects of the local culture.	Both indigenous and scientific names are used in local research projects, with results compiled in educational posters, booklets and videos.	Integration of social and cultural analysis and agrarian and environmental studies for resource use.
22	Support of Indigenous Knowledge for the Use and Conservation of Biological Diversity of Ethnic Minorities in Three Ecological Regions in Yunnan, Southwest China	Yunnan, China	Conservation of food plants, medicinal plants as well as livestock is promoted and augmented by the use of local knowledge.	Protection and promotion of traditional and indigenous knowledge. The project partners with ethnic minorities as well as forest and conservation agencies and documents traditional knowledge for the conservation and sustainable use of biodiversity.	Video and other visual activities support indigenous language use.	Use of traditional knowledge for conservation of plants and animals.
23	Tado Cultural Ecology Conservation Program	East Nusa Tenggara, Indonesia	Researching, documenting, archiving and restoring/reviving biodiversity of the Tado and Waerebo people and their ancestral lands. The focus is on conserving both native taxa and native traditions.	Tado and Waerebo communities with support of an international NGO have researched and documented about 600 traditional ethnobotanical practices (utilitarian, medicinal, social, decorative, ritual and narrative uses) associated with 200 species of plants. Revival of sacred and secular rituals; traditional varieties of rice conserved; ancestral lands mapped.	Project documents describing cultural practices are published bilingually in Kempo Manggarai (the language of the Tado, spoken by 18 villages in the Manggarai region) and Bahasa Indonesia. Several hundred traditional Kempo Manggarai sayings and dozens of traditional recipes (involving native plants and insects) have been documented.	Conservation of biological and cultural heritage by documenting traditional uses of plants, traditional foods and community-based ecotourism.

#	*Project name*	*Location*	*Biodiversity conservation*	*Cultural conservation*	*Language revitalization*	*Linkages made*
24	Territorial Management in Brazil's Xingu Indigenous Park	Brazil	Natural resources management plan for Brazil's Xingu Indigenous Park that includes biodiversity conservation based on traditional knowledge.	Ethnographic, cultural mapping (traditional territories, sacred sites, fishing and hunting locales) of over 7 million acres of savannah and lowland tropical rainforest in Brazil's Xingu Indigenous Park will drive the conservation of biodiversity in the Xingu region.	Not a project focus.	Project establishes a culturally appropriate management scheme for the park and its inhabitants.
25	The Significance of Non-Timber Forest Products Utilization and Cultural Practices in Rural and Urban Households: Implications for Biocultural Diversity	Eastern Cape, South Africa	Project promotes the conservation of wild resources, by studying and promoting their cultural value instead of utilitarian value.	Cultural value of wild resources is emphasized: project studies and promotes the importance of biodiversity for the cultural fabric of local communities.	Wild resources used in the study area are identified with their vernacular names.	Cultural practices are threatened by loss of biodiversity. Utilitarian and cultural values of wild resources are promoted in an integrated manner with biodiversity conservation strategies.
26	The Use of Aboriginal Traditional Knowledge in Species Assessment: A Case Study of Northern Canada Wolverines	Nunavut, Northwest Territories and Yukon, Canada	Committee on the Status of Endangered Wildlife in Canada (COSEWIC), responsible for evaluating the status of species in Canada, is concerned with the decreasing abundance of the wolverine.	Project focus is primarily the gathering of factual aboriginal traditional knowledge (rather than the cultural context of this knowledge), which is described and utilized in COSEWIC's species assessment.	Project is not language oriented.	Aboriginal traditional knowledge contributes to a federal regulatory process to protect endangered species.
27	Traditional Ecological Knowledge Relating to Marine Environment and Fishing on Lihir	New Ireland Province, Papua New Guinea	Project promotes sustainable management of threatened marine environments and fish stocks.	Traditional Lihirian fishing techniques, ownership and management of resources are documented and applied, as are restrictions on marine exploitation associated with ceremonies.	Teaching materials are developed in local languages.	Traditional management of resources used in management and conservation strategies.

#	Project name	Location	Biodiversity conservation	Cultural conservation	Language revitalization	Linkages made
28	Training Program of Indigenous Agro-Forestry Agents of Acre	Brazil	Project aims for ecological sustainability through environmental management of the indigenous territories, blending indigenous and modern technologies for resource management.	Indigenous Agro-Forestry Agents act as environmental educators to preserve and strengthen cultural diversity and enhance a sense of identity and social cohesion.	Agro-forestry agents are trained to write in the local first and second language. They each receive a bilingual and intercultural education based on linguistic acquisition and development in their own native languages as well as Portuguese.	Cultural practices and sense of identity are important for sustainable ecosystem management.
29	Whitefeather Forest Initiative	Ontario, Canada	Maintenance of forest cover, local biodiversity and the care of vulnerable species. The vision of the Pikangikum people honours the teachings and wisdom of the Pikangikum elders, which support the care and protection of the diversity of life and ecological richness of the land.	Project maintains the vitality and strength of the indigenous language, culture and knowledge tradition of the community: includes customary indigenous resource stewardship practices and management tools rooted in a rich indigenous knowledge tradition as well as in the spiritual and emotional connection to the land. Elders' statements on biodiversity are honoured and reflect the vast nature of their cultural horizon.	New economic and resource management context helps maintain the vitality and strength of the indigenous language, culture and knowledge tradition of the community. Project materials are made available in both Ojibwa and English.	Elders' goal is to develop new forest-based livelihood opportunities for the youth of Pikangikum in a context where the knowledge tradition, the language and stewardship values of the community including teachings of respect for biological diversity, play a leading role in guiding the development of the initiative.
30	Worlds of Difference	Ithaca, New York, US	Generating awareness of threatened biodiversity through a radio series focusing on threats to cultural diversity.	Radio programmes examine traditional cultural practices applied to new social problems in a period of rapid cultural change. Project includes an examination of the human role in sustaining wild and agricultural diversity.	Radio series includes pieces concerned with language revitalization efforts in various parts of the world.	Making linkages between socio-cultural change and ecological change.

#	*Project name*	*Location*	*Biodiversity conservation*	*Cultural conservation*	*Language revitalization*	*Linkages made*
31	Wik, Wik-Way and Kugu Ethnobiology Project	Queensland, Australia	Project addresses loss of plant-related local knowledge and practices and links to environmental degradation, e.g. loss of traditional burning regimes that are essential for maintaining habitats and conserving local biodiversity.	Project focus is on loss of traditional knowledge and how to sustain it for conservation of local biodiversity; includes resource management, fire management and use of wild species.	Wik and Kugu names of elements of their environment as well as local plant taxonomies and traditional land management techniques are documented.	Project makes explicit links between environmental degradation and loss of traditional knowledge, which in turn affects biodiversity.
32	Plant Resources: Traditional Knowledge of Irulars of Northern Tamil Nadu	Northern Tamil Nadu, India	Research examines the status of traditional knowledge of biodiversity so it can be revived and used for conservation as their forefathers did.	Research on local use of biodiversity for medicines, food, hunting, ceremonial purposes and relationship to spiritual beliefs. Gender dimension of knowledge is incorporated.	Project is not language oriented.	Research was conducted in order to understand the knowledge and the relationship of the Irular community with the local biodiversity in order to conserve it. Local knowledge of plant biodiversity used for medicines, food, hunting and ceremonial purposes acts to conserve biodiversity.
33	Local Level Ecosystem Assessment in India	India	Recording knowledge of local biodiversity and its uses contributes to a national database, the People's Biodiversity Register, mandated by national legislation in support of biodiversity conservation.	Project supports maintenance of local innovation, cultural practices such as sacred groves, sacred water bodies that have cultural meaning and constitute conservation practices.	Project documents species names in local languages for the purposes of linking to scientific nomenclature.	Linkages made by recognizing the belief systems that act to conserve forests and water bodies.
34	Medicinal Plants of Antiquity	Mediterranean, Europe	Historical knowledge and uses of plants may provide the rationale for future conservation.	Focus is on ancient knowledge of adaptation to the environment. Ancient knowledge of natural resources for medicines is being recovered.	Project is not language oriented.	Past human cultural adaptations to the environment may provide information on species importance and environmental adaptation.

#	*Project name*	*Location*	*Biodiversity conservation*	*Cultural conservation*	*Language revitalization*	*Linkages made*
35	Ethnocartography and Self-Demarcation of Indigenous Peoples' Lands in Venezuela as Tools for Biocultural Diversity Conservation	Venezuela	Project seeks to obtain exclusive rights to land occupation and use for local groups whose lifestyle and resource use practices are historically demonstrated to be compatible with environmental conservation.	Cultural-historical documents provide the basis for land claims. Traditional ethno-cartographical knowledge to be taught in schools. Recording and systematization of a database of local knowledge about the land, its natural resources, and associated sustainable use concepts and practices.	Taxonomic knowledge is formally encoded, organized and transmitted through language, and changes in language are being documented to examine links between language, traditional knowledge and environmental change.	Lifestyle and resource use practices are historically demonstrated to be compatible with environmental conservation. Project has also led to greater awareness of the value of traditional knowledge of the environment.
36	Weavers of Life (*Tejedores de Vida*)	Colombia	Project addresses conservation issues of one of the last stands of native Andean primary forest, located in the collective communal territory of the Muisca of Sesquilé. The species and important water resources are threatened by economic interests as well as by natural phenomena such as fires.	Project strengthens sustainable cultural and subsistence practices, including the recovery of traditional cultivars, languages and activities such as weaving, traditional medicine and pottery. The project seeks to revitalize for the new generations the cultural heritage and the ancestral cosmological knowledge that would otherwise be destined to disappear.	Muisca language has almost disappeared due to colonization and the marginalization of the culture. The project is reintroducing Muisca words in the social and subsistence practices of the community. A goal is to teach Muisca language in the schools.	Project aims to revitalize and strengthen cultural heritage for the management of local biodiversity in the territory.
37	Community-Based Documentation of Indigenous Knowledge, Awareness and Conservation of Cultural and Genetic Diversity of Bottle Gourd (*Lagenaria siceraria*) in Kitui District in Kenya	Kenya	Agrobiodiversity conservation of over 50 landraces of bottle gourd (*Lagenaria siceraria*). Reintroduction of lost landraces and distribution of seed to farmers.	Project documents and revitalizes the high cultural significance of the bottle gourd, which has been cultivated for 10,000 years.	About 70 names describing the bottle gourd landraces in local language and associated songs and proverbs are being documented. Knowledge documented on audiotapes or published in journals or a national database in the community's own language.	Conservation of the cultural value of the high biodiversity of the bottle gourd.

#	*Project name*	*Location*	*Biodiversity conservation*	*Cultural conservation*	*Language revitalization*	*Linkages made*
38	Transforming the Cage	British Columbia, Canada	Project focus is on the link between healthy livelihoods and cultural revival through cultural centres, retreats and programmes. Project seeks to restore connection to the environment that is an intimate part of cultural practices for spiritual balance.	Project develops strategies to enable ownership of ancestral teachings as a relevant and effective vehicle for positive change. It provides guidance in reciprocal relationships with the environment and in creating spiritual balance in the individual and collective.	Project philosophy is that language is key in returning to the ancestral teachings and the guiding force for change.	Ancestral teachings based on cultural practices guide the relationship to the natural environment.
39	*Ndee bini' bida'ilzaahi*: Pictures of Apache Land	Arizona, US	Ecological restoration of diverse wetland communities; instilling in the youth a commitment to restoration of their land and waters.	Project builds on elders' knowledge of plants in culturally important sites of the past as a basis for restoration.	Apache names for plants, places and other ecological features are emphasized. Participants have demonstrated deeper cultural knowledge, including a greater willingness and proficiency to speak in Apache.	Culturally and ecologically important sites are being restored.
40	Vanishing Voices of the Great Andamanese	Andaman Islands, India	Ecological knowledge of the flora and fauna, names and uses of medicinal plants and terms related to hunting and gathering form a major part of a trilingual Andamanese dictionary.	Project collects oral texts, writes sociolinguistic sketches and makes extensive audio and video recordings of the surviving 50 Great Andamanese.	Trilingual dictionary (Great Andamanese–Hindi–English) is documenting the Great Andamanese language that is spoken fluently by only seven people.	Highlights knowledge of local ecology of the Andaman environment (such as cues in the environment that preceded the 2005 tsunami), contained in the nearly extinct Andamanese language and way of life.
41	Knowledge and Language Revitalization in Hawaii	Hawaii	Not a direct focus on biodiversity, but educational projects offer classes in traditional farming, medicinal herbs and gathering of native forest products, traditional fishing and aquaculture to celebrate and record the history of the Hawaiian people. Teaching incorporates Hawaiian models of land stewardship and caring.	Curriculum is grounded in a native perspective that makes connections to mainstream academics through indigenous approaches to learning; traditional songs, dances and culturally based practices are being documented.	Language is the main focus of the project, through education in the Hawaiian language from pre-school through to university.	Culturally based environmental knowledge is revitalized through the reinstitution of language and traditional teaching and learning techniques.

#	Project name	Location	Biodiversity conservation	Cultural conservation	Language revitalization	Linkages made
42	Mapping Aboriginal Cultural Values in the Wet Tropics World Heritage Area	Queensland, Australia	Management and protection of high biodiversity of Northeastern Australia's wet tropical forests to be assisted by protection and management of Aboriginal cultural heritage and values.	Local Aboriginal peoples (younger and older) participate in recording beliefs, knowledge, heritage and practices for collaborative management of Wet Tropics World Heritage Area.	Language training programmes will address loss of Aboriginal languages in the area. Aboriginal languages can be expected to form an important part of the case for renominating and relisting the area as a biocultural landscape.	Generations of Aboriginal knowledge, values and practice must be taken into account for the biodiversity of the biocultural landscape to be protected and managed.
43	A Collaborative Social and Biological Study with Gamo Elders of the Importance for Biocultural Diversity of Living Indigenous Sacred Sites in the Gamo Montagnard Region of Southwest Ethiopia	Southwest Ethiopia	High diversity of living indigenous sacred sites in four districts has been protected for generations but is now threatened by changes in the value system. Nursery sites established in some communities to help restore degraded sacred forests.	Project seeks recognition of the importance of culturally based beliefs and traditional institutions that have maintained sacred forests. There has been support for people to undertake their ritual festivals. People are gratified that their indigenous religion is coming out into the open after 30 years of suppression.	Project is not language oriented.	Culturally based beliefs and the system of traditional institutions that conserve sacred forests are considered vital to their continued maintenance. The project aims to raise awareness among the Gamo elders as well as in government of the importance of culturally based conservation.
44	Talking the Walk: Language as the Missing Ingredient of Biodiversity Conservation? An Investigation of Plant Knowledge in the West Usambara Mountains, Tanzania	West Usambara mountains, Tanzania	Not a direct focus on biodiversity, but project points to importance of maintaining local languages and traditional knowledge for conservation of local biodiversity.	Project stresses role of local languages and the environmental knowledge they embody for both cultural and environmental sustainability.	Project clarifies biocultural dynamics of language and mechanisms of language shift, and implications for language maintenance and biodiversity conservation.	Project explores the interconnectedness and interdependency between biological, cultural and linguistic diversity, widening the knowledge base of biocultural diversity theory in an African context.

#	*Project name*	*Location*	*Biodiversity conservation*	*Cultural conservation*	*Language revitalization*	*Linkages made*
45	Ethnobotany of Indigenous People of the Southern Rift Valley and Southwestern Ethiopia	Southwest Ethiopia	Botanical diversity is being documented by recording the extensive ethnobotanical knowledge of the indigenous peoples of the region. High biodiversity is essential to local livelihoods.	Traditional spiritual values have influenced people's behaviour toward the forests, and have played a role in protecting them and ensuring that some of the culturally valued trees and other medicinal plants are sustained.	Documentation of local plant names in the local languages will help preserve minority languages.	Biodiversity conservation through traditional uses of plants is being documented and encouraged.

Table A1.2 *Level of participation and means of knowledge transmission in projects; contribution to biocultural diversity policy*

#	*Project name*	*Location*	*Level of participation*	*Methods/institutions for knowledge/language transmission*	*Biocultural diversity policy*
1	A Review of the Birds and Plants of Bikini Atoll, Trees of the Marshall Islands and Fish of Micronesia	Majuro, Marshall Islands	No specific information provided.	Production of guides to local flora and fauna with species names in Marshallese, to be made widely accessible.	No specific information provided.
2	Indigenous Theory for Health: Enhancing Traditional-Based Indigenous Health Services in Vancouver	British Columbia, Canada	Project is based on ideology of First Nations, developed from the informal recommendations of traditional indigenous practitioners and by a traditional research group, supported by university and government funding.	Indigenous worldviews are strengthened through use and application to new issues. Reliance on a common language (English patois) is thought to assist in continued transmission of traditional principles and practices outside of specific language groups.	Project to assist negotiations for traditional health services with provincial and federal governments. Principle of holistic healing calls for protection of traditional medicine sites and sacred sites.
3	Jande Myra Ta Ka'a Rupi Ha (Our Trees of the High Forest): Ka'apor Ethnodendrology	Eastern Amazonian Brazil	Project works in close collaboration with a Ka'apor non-profit corporation.	No specific information provided.	No specific information provided.
4	Bamenda Highlands Forest Project	Northwest Province, Cameroon	Collaborative partnership between local communities and international conservation community based on common interests and certain agreed upon objectives.	Maintenance of traditional governance structures combined with creation of forest management institutions, which manage forests at the village or village group level. Umbrella groups and local NGOs also involved in supporting local forest management.	Cameroon government supports the project due to its obligations as a signatory to the CBD.

#	*Project name*	*Location*	*Level of participation*	*Methods/institutions for knowledge/ language transmission*	*Biocultural diversity policy*
5	Biocultural Diversity: Elaborating Theoretical Issues for Communities and Policy Makers	Victoria and Northern Territory, Australia	No specific information provided.	Development of databases that capture the underlying meaning of traditional worldviews.	Project works at the interface between academics and engagement in policy formulation and activism for indigenous peoples' rights.
6	Eco-cultural Health in the Sierra Tarahumara, Mexico	Sierra Tarahumara, Chihuahua, Mexico	Fully collaborative, with the community taking the lead in how the project proceeds based on their priorities.	Alternative education on Rarámuri language, culture and traditional knowledge, within an eco-cultural framework, is a key goal.	Project not policy oriented, but seeks to build Rarámuri capacity in relation to land tenure issues.
7	Collection and Documentation of Traditional Conservation Sites	Marshall Islands	Project calls for more participation of people living near new MPAs, as their use of resources is affected, as well as for more assistance from government to manage traditional conserved areas.	Documentation of traditional knowledge and beliefs linked to traditional conservation sites and other traditionally taboo areas in the Marshall Islands, and integration of traditional practices into legislation.	The project aims to integrate traditional concepts of conservation into the National Biodiversity Strategy and Action Plan.
8	Conservation in Managed Indigenous Areas (Conservación en Áreas Indígenas Manejadas, CAIMAN)	Ecuador	USAID-financed team in consultation with indigenous groups; indigenous peoples are fully integrated in the development and implementation of work plans.	Maintenance of certain key cultural elements (language and medicinal plant use) in order to resist massive cultural change.	Project supports indigenous federations in gaining legal rights to ancestral territories within Ecuadorian constitution and legal framework.
9	Crocodile Rehabilitation, Observance and Conservation	Northeast Luzon, Philippines	Contractual arrangements are made to formalize responsibilities of the partners in community-based conservation. In the contemporary political situation in the Philippine uplands, the full support of the local people is a necessity and the project links indigenous and local governments and the international conservation movement.	Project documents and revives traditional knowledge and practices that were beneficial for crocodile conservation, and promotes past traditional practices in a contemporary context.	Project assists indigenous groups in obtaining land rights and enshrines cultural traditions in law. Local municipality has become partner in conservation, enacting ordinances to protect crocodiles and establishing first crocodile sanctuary in the country, co-managed by local communities.

#	*Project name*	*Location*	*Level of participation*	*Methods/institutions for knowledge/language transmission*	*Biocultural diversity policy*
10	Dance for the Earth and for Her Peoples	Latin America, Caribbean, Africa and Europe	Diverse international group of professionals, who have contacts and work with community groups. Local communities have now taken on the idea themselves.	Project promotes the use of dance as a vehicle for the expression of cultural traditions.	Project is influenced and endorsed by IUCN's Commission on Environment, Economic and Social Policy (CEESP) and World Protected Areas (WCPA).
11	Environmental Applications Reference Thesaurus (EARTh)	Italy	Not applicable at current stage of project.	Not applicable at current stage of project.	Thesaurus is meant as tool for management and retrieval of information relevant to environmental research and policy; aims to enhance awareness among policy makers of cultural dimension of knowledge.
12	Establishing Marine Protected Areas and Spatio-temporal Refugia in the Roviana and Vona Vona Lagoons, Solomon Islands	Solomon Islands	Local communities are partners in a co-management arrangement including officials at local, regional and national levels. Workshops designed to encourage local participation.	Maintenance of knowledge and practices through implementation of customary tenure and practices in the present-day context. Project has helped establish the institutional infrastructure to sustain the protected areas.	Project seeks to legalize all Marine Protected Areas at provincial and national levels, based on co-management regime with local, provincial and national governments. Project is working to establish a set of guidelines for implementing marine conservation initiatives in the region.
13	Forests and Oceans for the Future	British Columbia, Canada	Project based on collaborative framework between university and community. Community members are active participants in all phases of the project.	Project facilitates the use of customary forms of governance for sustainability.	Project focus is on use of Gitxaała traditional ecological knowledge for provincial government land use planning. Project research is incorporated into the British Columbia government's Land Resource Planning Process.
14	Gwich'in Place Names and Traditional Land Use Project	Northwest Territories and Yukon, Canada	Gwich'in Social and Cultural Institute, the cultural and heritage arm of the Gwich'in Tribal Council, carries out project in collaboration with Gwich'in communities in the land claim area.	Project emphasizes bringing elders and youth together on the land for knowledge transmission. Development of resources (land-based history book, ethnobotany book, etc.) for schools and museums at both local and national levels, as well as a website that features a 'talking place name map' and virtual tours of the Mackenzie, Peel and Tsiigehtchic Rivers.	Project information is incorporated into the Gwich'in Land Use Plan and used to evaluate proposed land use activities in the Gwich'in Settlement Region. A Gwich'in Traditional Knowledge Policy guides all traditional knowledge research in the Gwich'in Settlement Region.

#	*Project name*	*Location*	*Level of participation*	*Methods/institutions for knowledge/ language transmission*	*Biocultural diversity policy*
15	Andean Project for Peasant Technologies (Proyecto Andino para las Tecnologías Campesinas, PRATEC)	Peru	Philosophy of *campesino* communities (belief that it is affection for the seeds and respect for all entities in the world that conserves diversity) leads the project.	Traditional practices such as the agro-festive cycle are promoted as a means of knowledge maintenance. Project documents traditional practices for an alternative to Western education curriculum so the knowledge is maintained with the youth in the school system.	Project focuses on policies to promote in-situ conservation of agrobiodiversity. Incorporation of local knowledge into school curriculum and adoption of local agricultural calendar have become national policy.
16	Jaru Ethnobiological Language Knowledge Project	Western Australia	The Kimberley Language Resource Centre is governed by an elected board of 12 Aboriginal directors accountable to a membership representative of the approximately 30 languages.	Knowledge and language transmission is through audiovisual and written resources, a DVD of traditional knowledge of trees in Jaru language, women's medicinal knowledge documented and bush trips for children.	Not policy oriented.
17	Linking Crop Diversity with Food Traditions and Food Security in the Hills of Nepal	Kaski, Nepal	Participatory research tools, with external control by researcher to meet academic objectives.	Affirmation of traditional practices contributes to maintenance and transmission of knowledge and practices.	Project influenced policy guidelines on tourism training centres concerning promotion of traditional foods.
18	Participatory Genetic Improvement of Traditional Crops and Native Tree Species	Costa Rica	Small NGO worked together with the family farming community to revive traditional seed diversity.	Encouraging seed exchanges helps maintain knowledge of agrobiodiversity. Youth are actively involved and information is included in studies at local university.	Costa Rican Ministry of Culture and Ministry of Health are collaborating in nationwide project to protect food traditions and sub-utilized foods, as food security is now seen as an issue of national security.
19	Promotion of Traditional Medicine and Indigenous Cultural Research and African Spirituality	Kampala District, Uganda	Local healers direct project and share their knowledge with government and academic specialists.	Project documents traditional practices and shares information via the Healing and Cultural Demonstration Institute.	No specific information provided.

#	*Project name*	*Location*	*Level of participation*	*Methods/institutions for knowledge/ language transmission*	*Biocultural diversity policy*
20	Support Project for the Ngäbe Indigenous People (Proyecto de Apoyo al Pueblo Indígena Ngäbe)	Costa Rica	Project is based on co-management principles and local priorities were the starting point for the project. There was an overall participatory process along with specific ones for individual activities (traditional story book, medicinal plants compilation, healers' apprentices, legal study, etc.).	Ngäbe youth are documenting oral history and songs, thus transmitting traditional culture between generations in school curricula.	Project produced a guidebook that interprets the legal rights of the Ngäbe to defend their territory and resources and claim their land rights. Project also includes a legal study to influence policy change with regard to indigenous rights to manage natural resources.
21	Social, Environmental and Economic Sustainability in the Context of Melanesian Mining Projects	New Ireland Province, Papua New Guinea	Collaborative approaches to the development of awareness and understanding of environmental impacts associated with mining.	Project has a local educational component whereby schools participate in various research projects, the results are compiled and the findings and photographs are presented in posters, booklets and videos that are made available to the schools.	No specific information provided.
22	Support of Indigenous Knowledge for the Use and Conservation of Biological Diversity of Ethnic Minorities in Three Ecological Regions in Yunnan, Southwest China	Yunnan, China	Project is based on community-driven participatory action research. It links institutions and organizations with representatives of ethnic minorities and conservation agencies in a cooperative agreement.	Use of visual materials thought to enhance awareness and understanding of the value of indigenous knowledge. Project emphasizes institutional strengthening and is establishing local and regional networks for the exchange of experiences among the pilot areas through local seed fairs, cross-farm visits and study tours.	Yunnan Initiative that guides project is based on Declaration of Belém, Kunming Action Plan and International Society of Ethnobiology's Code of Ethics. It also endorses CBD's call for respect of cultural and spiritual values for sustainable development.

#	*Project name*	*Location*	*Level of participation*	*Methods/institutions for knowledge/ language transmission*	*Biocultural diversity policy*
23	Tado Cultural Ecology Conservation Program	East Nusa Tenggara, Indonesia	Project is dedicated to collaborative research. Tado community has run programme for several years, with financial, administrative, logistical and technical support from an international NGO. Tado and Waerebo research associates collectively administer research programmes in their own communities.	Project focus is on documenting and reviving traditional knowledge and practices for use in conservation. Research results are systematically disseminated to the community in written documents and readings enabling elders and community members to regularly review, critique and augment results.	Collaboration between Tado community and international NGO follows tenets of CBD and UN-WGIP Principles and Guidelines for the Protection of the Heritage of Indigenous Peoples regarding the sharing of benefits and responsibilities for the conservation of biocultural diversity. Programme embodies principles of ISE's Code of Ethics.
24	Territorial Management in Brazil's Xingu Indigenous Park	Brazil	14 tribal groups and National ministries formed a partnership. Supporting NGO passed on the assets of its regional field office to the leading indigenous organizations of the Xingu Indigenous Park, who were able to manage their own land and cultural conservation efforts.	Development of a 'life plan' and management scheme meant to serve as management guidelines.	Project worked with Brazilian indigenous affairs agency FUNAI to map Xingu Indigenous Park territory and develop culturally appropriate management plan for park. Project also sought to sensitize relevant public agencies to problems in Xingu region to foster defence of indigenous territories.
25	The significance of Non-Timber Forest Products Utilization and Cultural Practices in Rural and Urban Households: Implications for Biocultural Diversity	Eastern Cape, South Africa	No specific information provided.	Project aims to promote awareness of the cultural value of resources and of the links between cultural and biological diversity among students.	Not policy oriented.

#	*Project name*	*Location*	*Level of participation*	*Methods/institutions for knowledge/ language transmission*	*Biocultural diversity policy*
26	The Use of Aboriginal Traditional Knowledge in Species Assessment: A Case Study of Northern Canada Wolverines	Nunavut, Northwest Territories and Yukon, Canada	Project is based on consultative research.	Project not directly related to transmission within Aboriginal groups themselves, but promotes knowledge transmission via the encoding of aboriginal traditional knowledge in legislation.	Project worked on inclusion of aboriginal traditional knowledge in national-level assessments of species at risk conducted by Committee on the Status of Endangered Wildlife in Canada. Research results expected to markedly change governmental wildlife policies.
27	Traditional Ecological Knowledge relating to Marine Environment and Fishing on Lihir	New Ireland Province, Papua New Guinea	Project is a collaboration between Lihir communities and two Australian universities. Substantial buy-in from high school, community schools, and PNG National Education Department.	Use of local knowledge for local management strategies. Teaching materials for schools are generated from research.	Project attempts to influence local-level policy to reduce over-exploitation of fish stocks.
28	Training Program of Indigenous Agro-Forestry Agents of Acre	Brazil	Project focus is on training indigenous peoples who then work with indigenous communities.	Project integrates traditional knowledge with scientific technologies for application in natural resource management strategies.	Training programme responds to political demands from regional indigenous populations and includes awareness of environmental legislation and domestic policies related to demarcation of indigenous territories.
29	Whitefeather Forest Initiative	Ontario, Canada	Elders of the community take a leading role in planning through a steering group; youth work with elders. All partnerships are based on achieving respect through dialogue.	Elders teach and transmit knowledge to youth in an active programme.	Local knowledge played lead role in the policy framework for the Northern Boreal Initiative (NBI), which uses community-based land use planning. Project also seeks to establish a linked network of protected areas. A Protected Areas Accord was signed in 2002, with the goal of achieving UNESCO World Heritage status.

#	*Project name*	*Location*	*Level of participation*	*Methods/institutions for knowledge/ language transmission*	*Biocultural diversity policy*
30	Worlds of Difference	Ithaca, New York, US	Focus is on documentation based on consultation.	The use of media may help reaffirm the value of traditional knowledge and practices.	Not policy oriented, but contributed to awareness raising through media.
31	Wik, Wik-Way and Kugu Ethnobiology Project	Queensland, Australia	Cross-cultural collaborative project, between scientists and local experts. Elders' concern over the loss of traditional knowledge was the reason the project was initiated.	Database that integrates local Wik knowledge with scientific knowledge is used as an educational tool. Project focus is to provide tools to promote intergenerational transmission of local knowledge.	Wik, Wik-Way and Kugu Land and Sea Management Centre has policy of promoting biocultural diversity within the region, following ISE's Code of Ethics. Local knowledge has contributed to policy at the regional level and to a national oceans policy.
32	Plant Resources: Traditional Knowledge of Irulars of Northern Tamil Nadu	Northern Tamil Nadu, India	No specific information provided.	Project stressed importance of generating appreciation for cultural uses of plant in order to revive intergenerational transmission of ethnobotanical knowledge to younger generations.	Project sought to identify obstacles to greater community participation in government conservation planning and implementation.
33	Local Level Ecosystem Assessment in India	India	People's Biodiversity Register is mandated by national legislation, but awareness of intellectual property rights lets people decide what knowledge to contribute to the register.	Formalized database at the national level contributes to awareness of value of traditional knowledge.	Project is based on India's National Biological Diversity Act. Decentralization of ecosystem management allows for the use of traditional values and knowledge and a coordinated effort between local and national levels for biodiversity policy. Project works within the CBD mandate of Intellectual Property Rights and Access and Benefit Sharing related to traditional knowledge.
34	Medicinal Plants of Antiquity	Mediterranean, Europe	Not a field project involving community participation.	Research and documentation of past traditional knowledge contributes to knowledge transmission to the present day.	Not policy oriented.

#	*Project name*	*Location*	*Level of participation*	*Methods/institutions for knowledge/ language transmission*	*Biocultural diversity policy*
35	Ethnocartography and Self-Demarcation of Indigenous Peoples' Lands in Venezuela as Tools for Biocultural Diversity Conservation	Venezuela	Active collaboration between researchers and members of local communities, who are the principal data collectors and processors; researchers act as advisers and assist in data analysis and document preparation.	Project involves plans to develop teaching materials for local schools. Project has also led to greater conscious awareness of the value of traditional knowledge of the land and environment for the current and future lives of the people.	Cartographic demographic, and cultural-historical documents produced in order to support efforts to secure legal ownership and title to the land occupied by Hotï and Eñepa ethnic groups according to Venezuelan constitution.
36	Weavers of Life (*Tejedores de Vida*)	Colombia	Project arises from the initiative of the Muisca community.	Project seeks to revitalize for the new generations the cultural heritage and the ancestral cosmological knowledge that would otherwise be destined to disappear.	Community efforts led to legal recognition of Muisca of Sesquilé according to Colombian constitution. Regional government environmental organization supports community activities relevant to cultural affirmation and land management.
37	Community-Based Documentation of Indigenous Knowledge, Awareness and Conservation of Cultural and Genetic Diversity of Bottle Gourd (*Lagenaria siceraria*) in Kitui District in Kenya	Kenya	Collaborative learning with the community, allowing time for meaningful interaction between all project participants.	Documenting and disseminating songs, stories and knowledge of the bottle gourd helps affirm and teach cultural knowledge. Community-based resource centre established as education centre for school children and others. Group has also shared indigenous knowledge, experiences and seeds with others in the district via seed fairs, workshops, indigenous knowledge competitions and joint planting activities.	Project's approach is to empower traditional knowledge holders and recognize their contribution at the national and scientific level, as well as to foster recognition of local peoples' rights. Awareness of project concept – conservation of traditional crop diversity for community development – not yet widespread in policy contexts, but growing.
38	Transforming the Cage	British Columbia, Canada	Project is situated within First Nations ideology and practised within those communities.	Teaching of ancestral law is seen as a vehicle of transformation and as a practice that will maintain cultural traditions.	Ancestral law is being promoted within the Nisga'a Lisims Government as a vehicle for sustainable prosperity and self-reliance.

#	*Project name*	*Location*	*Level of participation*	*Methods/institutions for knowledge/ language transmission*	*Biocultural diversity policy*
39	*Ndee bini' bida'ilzaahi*: Pictures of Apache Land	Arizona, US	The Cibecue Community School initiated the project.	Project teaches youth in the community about traditional Apache values for the land. Produced computer database, audiovisual materials and exhibit for tribal museum.	In aftermath of largest fire in US southwest history, project engaged in monitoring of springs and rehabilitation/ stabilization activities, expanding scope of federal post-wildfire response effort to better address impacts to eco-cultural resources.
40	Vanishing Voices of the Great Andamanese	Andaman Islands, India	Project helps people understand the issues around disappearing language, so they became very willing to participate.	Dictionary and grammar books will assist future transmission of language and knowledge. Book of photographs, children's books, CD of traditional songs and folk stories also made.	Project highlights the need for policy to assist in the revitalization of threatened languages and cultures.
41	Knowledge and Language Revitalization in Hawaii	Hawaii	Highly participatory. Native Hawaiian institutions carry out the project initiatives, administration and implementation.	He Lani Ko Luna Community Based Learning Centre offers programmes and field sites for hands-on learning; K-12 immersion school offers curriculum in Hawaiian language and an indigenous paradigm.	No specific information provided.
42	Mapping Aboriginal Cultural Values in the Wet Tropics World Heritage Area	Queensland, Australia	Project partnerships developed with training and research institutions and natural/cultural resource management bodies. Aboriginal Natural Resources Management Plan is a blueprint that outlines, for all levels of government and the broader community, how to develop equitable partnerships with Aboriginal peoples to address a wide range of social, cultural, environmental and economic issues.	Information management systems as well as computer interfacing, storage and access needs and options for documenting traditional knowledge are being investigated.	Aboriginal Resource Management Plan raises national awareness of role of Traditional Owners in ecologically sustainable development of northern Australia; plan aims to increase opportunities for and involvement of indigenous peoples in local and regional resource management.

#	*Project name*	*Location*	*Level of participation*	*Methods/institutions for knowledge/ language transmission*	*Biocultural diversity policy*
43	A Collaborative Social and Biological Study with Gamo Elders of the Importance for Biocultural Diversity of Living Indigenous Sacred Sites in Four Districts of the Gamo Gofa Montagnard Region of Southwest Ethiopia	Southwest Ethiopia	At project planning stage, indigenous community participation was minimal. Innovative approach adopted during fieldwork led to indigenous peoples, research collaborators and professionals forming an equal partnership.	Identification and documentation of customary laws and belief systems around sacred sites and the traditional institutions supporting them. Sacred sites have been mapped and threats identified. Indigenous plant species and threatened species are identified and documented.	Project's focus on conservation potential of traditional belief system helped convince both national governments and local communities of the value of local traditions. Workshops given to decision makers on importance of sacred sites for culture and biodiversity conservation and to increase biocultural diversity awareness among decision makers. Project seeks to give legal backing to custodians of sacred sites.
44	Talking the Walk: Language as the Missing Ingredient of Biodiversity Conservation? An Investigation of Plant Knowledge in the West Usambara Mountains, Tanzania	West Usambara mountains, Tanzania	University-based research project, in collaboration with local cultural and eco-tourism group and local communities.	Three books with local stories in the two local languages will be published to aid local environmental and cultural conservation.	Project contributed to national and international debates on the use of mother tongue as the language of instruction, and pointed to importance of indigenous languages and knowledge for education and biodiversity policy in Tanzania. It also pointed to the need for institutions involved in biodiversity conservation to use intercultural and multilingual practices.
45	Ethnobotany of Indigenous People of the Southern Rift Valley and Southwestern Ethiopia	Southwest Ethiopia	Project was initiated by academic staff at Addis Ababa University in collaboration with indigenous groups. It is establishing best practices for working together with indigenous peoples, based on mutual trust and equal participation for the fair and equitable sharing of benefits accrued, in line with the principles espoused in the CBD.	Project is introducing mechanisms of horizontal exchange of knowledge, experience and values to neighbouring areas and socio-cultural/ethnolinguistic groups. Project is raising public awareness of the values of the biodiversity.	Project aims to raise public awareness of the values of biodiversity and knowledge of resources and to introduce access and benefit sharing scheme based on principles of CBD.

Appendix 2
Survey Details

Survey procedures

The development of this sourcebook began in early 2004. The first phase of the work included the following tasks:

- elaborating project selection criteria and areas of emphasis;
- developing the survey tools;
- identifying survey dissemination channels and distributing the survey tools;
- establishing a database for the storage of survey data, bibliographic materials and other relevant information, and devising a data processing procedure.

Project selection criteria and areas of emphasis

These are described in Chapter 3.

Development of survey tools

In order to search for biocultural diversity conservation projects, programmes and initiatives worldwide, a short questionnaire, or survey form, was made available in English, Spanish, Portuguese, French and Russian (see English version below). The form was designed to record initial project details: project name, supporting organization, location, contact information; a brief narrative description of the project and of the project's contributions to biocultural diversity conservation; a set of keywords by which to identify the project; and the project's main area(s) of emphasis, with a description of how the project addressed the area(s) in question. The purpose of this tool was to obtain preliminary information about projects for potential inclusion in the sourcebook, with additional questions to follow after an initial review of survey responses.

The survey form was accompanied by a call for information for direct email distribution to potential contributors, which contained a general description of the project, its rationale and its criteria (see below), and by an expanded two-page description for those interested in participating. Another version of the call for information was designed for dissemination through journals, newsletters, electronic lists and websites. All versions of the survey form were also made available for downloading on Terralingua's website, along with the other survey materials. Respondents to the survey were later sent an additional questionnaire with a more detailed set of questions (see below).

The call for information stated that, in compiling information for the sourcebook, Terralingua would follow established ethical criteria of information gathering and dissemination (free prior informed consent, right of veto, right to decide which information should or should not be made public, right to intellectual property over information). The document specified that ethical conduct would be an ongoing focus throughout our collaboration, and that arrangements would always be subject to renegotiation if new concerns arose. In the course of the survey, all contributors readily agreed not only to the circulation of materials within the survey group, but also to disseminating the information to a larger public through print and web media.

Survey dissemination

An extensive contact list for dissemination of the call for information was compiled drawing from Terralingua's worldwide network as well as from internet searches and other sources. This list included a variety of parties potentially interested in contributing to and/or publicizing the survey:

- individuals;
- organizations;
- journals, newsletters, bulletins, electronic lists and websites;
- indigenous networks;
- professional networks in the fields of conservation, cultural survival and language endangerment/maintenance/revitalization;
- professional societies in the same fields.

While the survey could not be expected to be all-inclusive, the goal was to find as many examples as possible of projects with the intended profile and with a geographic spread spanning all continents. For this purpose, the call for information included an invitation for recipients to pass the materials on to interested others. The intent of this 'snowball' method was to reach out to grassroots projects beyond Terralingua's existing network and beyond electronic and print channels.

Database development and data processing procedure

For survey purposes, Terralingua adopted the free ICONS database, which uses the Microsoft Access platform. ICONS is a powerful tool for storing and cataloguing data and for creating cross-referenced databases of organizations, programmes, bibliographies and other related information. A customized version of ICONS was created for the survey, with three functions:

- storing information about individuals and organizations involved in biocultural diversity projects;
- tracking Terralingua's correspondence with those individuals and organizations;
- storing the information from completed surveys in order to categorize and analyse it, and to generate reports.

The database has two main integrated modules: contacts and projects. The contacts module:

- currently has 820 entries, including both those from the original contact list mentioned above and further contact names acquired over the survey period;
- stores source, type of contact (organization or person, or both), contact names and affiliations, contact details (address, phone and fax numbers, and emails), project affiliation and contact activity information (date contacted and by whom).

The projects module:

- currently has 45 entries, each coded by means of a unique identifying number;
- stores all information (parallel fields) received from the surveys;
- is fully cross-referenced to the contacts module.

Survey tools

Call for information

Box A2.1 contains the text of the cover letter sent to potential sourcebook contributors.

Box A2.1 Cover Letter to Potential Sourcebook Contributors

Date

Dear Colleague,

Terralingua (www.terralingua.org) is undertaking the compilation of a Global Sourcebook on Biocultural Diversity and is seeking information on projects that are effectively bridging cultural and biological diversity issues. The purpose of this letter is to invite your input in a survey of biocultural diversity projects, programs, and initiatives.

'Biocultural diversity' refers to the linkages and interdependence between biological, cultural and linguistic diversity. All too often, languages, cultural practices and indigenous knowledge are eroding due to global change, resulting in a breakdown in the human–environment relationship. This breakdown underlies many of the environmental and social problems humanity is facing. Thus, the sustainability of 'natural' environments goes hand in hand with the sustainability of the associated human communities, and vice versa.

The promotion and development of this approach will benefit from collaboration among those involved in research and on-the-ground projects that are biocultural in nature. In developing the Global Sourcebook on Biocultural Diversity, supported by The Christensen Fund, Terralingua would like to work in partnership with biocultural diversity project participants to give this field its very first global source of information. Based on the survey, we anticipate further collaboration and information gathering, in order to select some projects as 'model' examples that support biocultural diversity, and which specifically highlight local stories in the voices of the people involved. Results of this joint venture in establishing a new network of people involved in biocultural diversity projects will be made widely available through various channels, both in print and in electronic form. All collaborators will also receive a copy of the Sourcebook for their own reference.

Benefits of this collaboration will be to local communities, non-governmental organizations, policy makers, governments, funders, researchers, media, and the general public. Specific benefits to project participants from working together with Terralingua to compile information for the Sourcebook, include increasing the visibility of biocultural diversity projects and the development of a network of people actively involved in biocultural diversity conservation through sharing information and experiences. Terralingua will support the creation of such a network and contribute to identifying avenues for advancing the network's shared goals.

We are asking you to read through and fill out the attached survey form, keeping in mind that Terralingua would like to emphasize the integration of cultural (including linguistic) and biological diversity conservation. We are seeking research or applied projects or those aspects of projects that recognize the essential link between local language, knowledge and the environment in the design of equitable and sustainable solutions to environmental and social problems. We are also seeking projects that are initiated and conducted by local beneficiaries themselves, or else jointly planned, led, and managed by both local people and outsiders. In compiling information for the Sourcebook, Terralingua will follow established ethics of information gathering and dissemination. This will be an ongoing process throughout our collaboration and arrangements will always be subject to re-negotiation if new concerns arise.

We look forward to hearing from you with regard to your project(s) and how it is/they are furthering the goals of global biocultural diversity conservation. If you know of other people involved in biocultural diversity projects, please send this letter and survey form on to them – we are most interested in reaching small, locally based projects. If you have any suggestions, comments or questions about the Sourcebook or the attached survey form, please do not hesitate to get in touch with us at Terralingua, at any of the contacts listed in the body of the email.

Ellen Woodley, Sourcebook Coordinator

Survey form

We here reproduce the survey form sent to potential sourcebook contributors (made available in English, French, Spanish, Portuguese and Russian):

Terralingua

Global Sourcebook on Biocultural Diversity Survey Form

Basic Project Details

Name of Project (please use full name and include any acronyms)

Name of Supporting Organization(s)

Project Location(s) (Village/Town/City; Province/State/Region; Country)

Start Date			End Date			Project Director(s) / Principal Investigator(s) [Name, Title]
mm	*dd*	yyyy	*mm*	*dd*	yyyy	

Project Contact Details

Address

Province/State/Region/Area	Postal or Zip Code	Country

URL (website address):

http://

Name of Contact Person #1

Phone # (including country and area code)

\+ Ext.

Fax # (including country and area code) | **E-mail address**

\+

Name of Contact Person #2

Phone # (including country and area code)

\+ Ext.

Fax # (including country and area code) | **E-mail address**

\+

The Sourcebook surveys research and applied projects that reflect four main areas of focus in biocultural diversity conservation (see next page) and the connections between them. Projects of interest analyze and/or contribute to supporting the links between ecological and socio-cultural resilience and sustainability. Emphasis will be placed on those projects that are initiated by or based on close collaboration with indigenous, minority, or local communities.

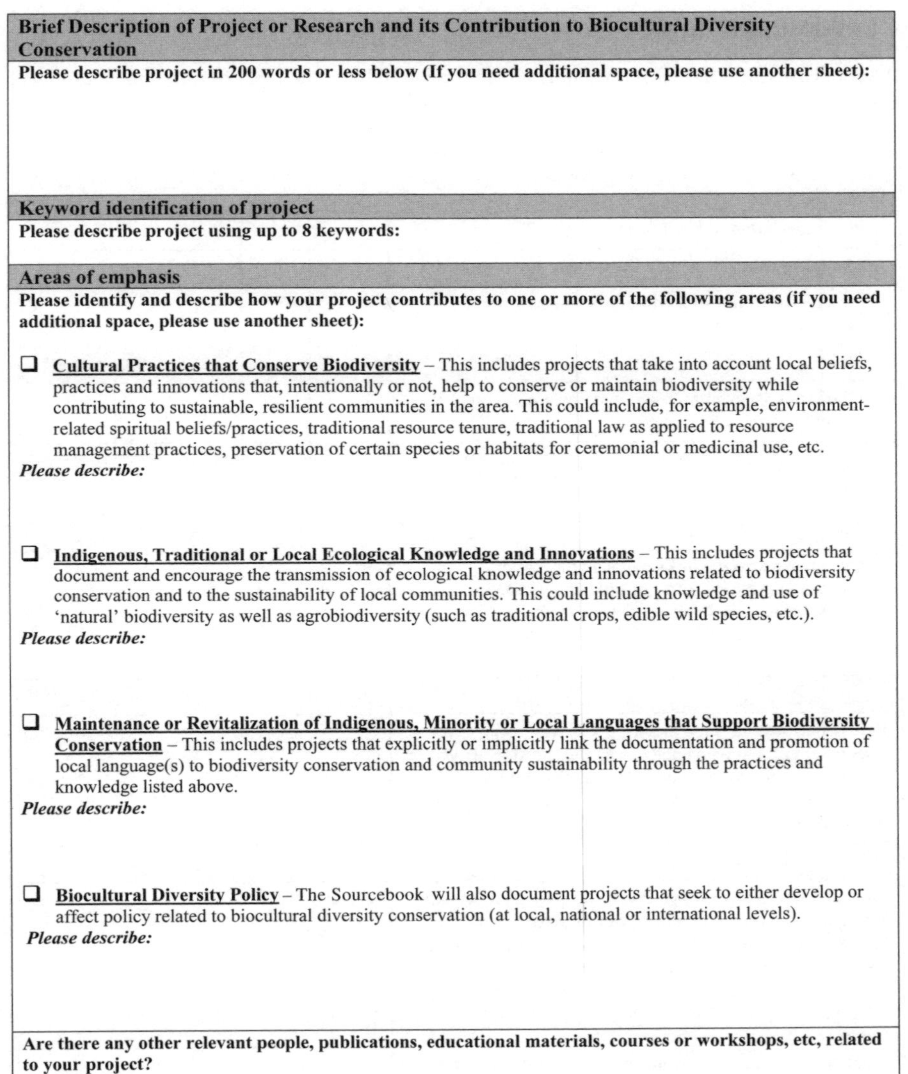

Brief Description of Project or Research and its Contribution to Biocultural Diversity Conservation

Please describe project in 200 words or less below (If you need additional space, please use another sheet):

Keyword identification of project

Please describe project using up to 8 keywords:

Areas of emphasis

Please identify and describe how your project contributes to one or more of the following areas (if you need additional space, please use another sheet):

❑ **Cultural Practices that Conserve Biodiversity** – This includes projects that take into account local beliefs, practices and innovations that, intentionally or not, help to conserve or maintain biodiversity while contributing to sustainable, resilient communities in the area. This could include, for example, environment-related spiritual beliefs/practices, traditional resource tenure, traditional law as applied to resource management practices, preservation of certain species or habitats for ceremonial or medicinal use, etc.

Please describe:

❑ **Indigenous, Traditional or Local Ecological Knowledge and Innovations** – This includes projects that document and encourage the transmission of ecological knowledge and innovations related to biodiversity conservation and to the sustainability of local communities. This could include knowledge and use of 'natural' biodiversity as well as agrobiodiversity (such as traditional crops, edible wild species, etc.).

Please describe:

❑ **Maintenance or Revitalization of Indigenous, Minority or Local Languages that Support Biodiversity Conservation** – This includes projects that explicitly or implicitly link the documentation and promotion of local language(s) to biodiversity conservation and community sustainability through the practices and knowledge listed above.

Please describe:

❑ **Biocultural Diversity Policy** – The Sourcebook will also document projects that seek to either develop or affect policy related to biocultural diversity conservation (at local, national or international levels).

Please describe:

Are there any other relevant people, publications, educational materials, courses or workshops, etc, related to your project?

Please list (If necessary, use another sheet):

Questionnaire table

Respondents to the initial sourcebook survey form were sent the table of questions reproduced here.

Pertaining to the community or region that the project is based in:	Yes	No	Reason(s)
Is local **knowledge** of biodiversity being lost?			
Are **languages** being lost (are there fewer speakers now)?			
Are traditional **practices and beliefs** related to local biodiversity being lost?			
			Explanation/Comment
Are the losses listed above seen as a problem by community members?			
			List
Are there specific threats to **biodiversity**?			
	List		
Can you make **specific links** between the loss of biodiversity and the loss of associated knowledge local practices, beliefs, and languages? For example as languages decline and/or practices and beliefs are lost, does this affect the local conservation of biodiversity (and vice versa)?			
	Describe		
How **collaborative** is the project – who was involved in the early design phase; who implements and monitors it? Would you say that this was a project that was the initiative of the local/Indigenous communities involved or someone outside these communities?			
	List		
What would your co-workers and other project participants like to see coming out of a **network** of these biocultural diversity projects that Terralingua is attempting to facilitate (i.e. increased visibility, funding opportunities, opportunity to learn from and share with like minded people, etc.)			
	List		
What are the **most important** things to share with others about this project? For example, what has worked, what needs to be improved, what lessons have been learned?			
	List		
What are the main **successes** of the project?			

Appendix 3

Survey Contributor Information

Table A3.1 *Contact information for project contributors by project name and descriptive title*

#	*Project names as given by contributors*	*Project descriptive titles as given in Chapter 4*	*Contact information for contributors and projects*
1	A Review of the Birds and Plants of Bikini Atoll, Trees of the Marshall Islands and Fish of Micronesia	Reconnecting with Natural and Cultural Heritage: Flora and Fauna of the Marshall Islands	Nancy Vander Velde nancyv@ntamar.net Jorelik Tibon jortibon@ntamar.net
2	Indigenous Theory for Health: Enhancing Traditional-Based Indigenous Health Services in Vancouver	Supporting Traditional Health Practices in Urban Areas: Indigenous Theory for First Nations Health in Canada	Dawn Marsden dmarsden@naho.ca
3	Jande Myra Ta Ka'a Rupi Ha (Our Trees of the High Forest): Ka'apor Ethnodendrology	Protection of an Indigenous Reserve: the Ka'apor People of Amazonian Brazil	William Balée wbalee@tulane.edu
4	Bamenda Highlands Forest Project	Taking Conservation into Our Own Hands: Forest Protection and Management by Highland Communities in Cameroon	Jonathan Barnard Jonathan.Barnard@birdlife.org John DeMarco demarcojohnf@yahoo.ca www.birdlife.org/action/ground/bamenda/bamenda4.html
5	Biocultural Diversity: Elaborating Theoretical Issues for Communities and Policy Makers	Bridging the (Digital) Gap: Aboriginal and Scientific Knowledge of Biodiversity in Northern Australia	Helen Verran hrv@unimelb.edu.au David Turnbull turnbull@deakin.edu.au www.cdu.edu.au/centres/ik/ikhome.html
6	Eco-cultural Health in the Sierra Tarahumara, Mexico	Recovering Landscape Health and Cultural Resilience in the Sierra Tarahumara	David J. Rapport drapport@ecohealthconsulting.com www.terralingua.org/projects/Sierra/sierra.htm
7	Collection and Documentation of Traditional Conservation Sites	Taboos and Conservation: Traditional Conservation Sites in the Marshall Islands	Nancy Vander Velde nancyv@ntamar.net Jorelik Tibon jortibon@ntamar.net
8	Conservation in Managed Indigenous Areas (Conservación en Áreas Indígenas Manejadas, CAIMAN)	Protecting Territories and Biodiversity: Indigenous Capacity Building in Ecuador	João de Queiroz joao.dequeiroz@sur.iucn.org

#	*Project names as given by contributors*	*Project descriptive titles as given in Chapter 4*	*Contact information for contributors and projects*
9	Crocodile Rehabilitation, Observance and Conservation	Life with Crocodiles: Reintroducing Human-Wildlife Coexistence in the Philippines	Jan van der Ploeg vanderploeg@cml.leidenuniv.nl www.cvped.org/v3/croc.php
10	Dance for the Earth and for Her Peoples	Strengthening Culture and Conservation through Intangible Heritage and Performing Arts: The 'Dance for the Earth and for her Peoples' Initiative	Rob Wild robwild_2005@yahoo.co.uk
11	Environmental Applications Reference Thesaurus (EARTh)	The Language of the Environment: A Comparative Environmental Thesaurus	Fulvio Mazzocchi mazzocchi@iia.cnr.it http://uta.iia.cnr.it
12	Establishing Marine Protected Areas and Spatio-temporal Refugia in the Roviana and Vona Vona Lagoons, Solomon Islands	Integrating Customary Tenure Systems in Marine Protected Areas: A Solomon Islands Example	Shankar Aswani aswani@anth.ucsb.edu
13	Forests and Oceans for the Future	Traditional Knowledge for Sustainability: Land use Planning Among the Gitxaała of British Columbia, Canada	Charles Menzies cmenzies@interchange.ubc.ca www.ecoknow.ca
14	Gwich'in Place Names and Traditional Land Use	Working with Traditional Knowledge in Land Use Planning: Gwich'in Place Names, Land Uses, and Heritage Sites in the Northern Territories of Canada	Ingrid Kritsch Ingrid_Kritsch@learnnet.nt.ca www.gwichin.ca/TheGwichin/Gwichin.html
15	Andean Project for Peasant Technologies (Proyecto Andino para las Tecnologías Campesinas, PRATEC)	Promoting Cultural and Biological Diversity: An Educational Program for Rural Communities in Peru	Jorge Ishizawa jorge.ishizawa@gmail.com Grimaldo Rengifo pratec@ddm.com.pe www.pratec.org.pe
16	Jaru Ethnobiological Language Knowledge Project	Caring for Country: Transmission of Aboriginal Environmental Knowledge in Western Australia	Siobhan Casson ldo@klrc.org.au (contact person for Kimberley Language Resource Centre Aboriginal Corporation)
17	Linking Crop Diversity with Food Traditions and Food Security in the Hills of Nepal	Culturally Rich Agro-ecosystems: Maintaining Traditional Beliefs for Food Security in Nepal	Laxmi Pant laxmipant@hotmail.com
18	Participatory Genetic Improvement of Traditional Crops and Native Tree Species	Reviving Traditional Seed Exchange and Cultural Knowledge in Rural Costa Rica	Felipe Montoya Greenheck milpa99@gmail.com
19	Promotion of Traditional Medicine and Indigenous Cultural Research and African Spirituality	Promoting Traditional Medicine, Indigenous Cultural Research, and African Spirituality in Uganda	Sekagya Yahaya Hill PROMETRA UGANDA PO Box 16465 Kampala Uganda
20	Support Project for the Ngäbe Indigenous People (Proyecto de Apoyo al Pueblo Indígena Ngäbe)	Strengthening Indigenous Cultural Heritage through Capacity Building in Costa Rica	Hugh Govan hgovan@compuserve.com Rigoberto Carrera www.tuva.org

#	*Project names as given by contributors*	*Project descriptive titles as given in Chapter 4*	*Contact information for contributors and projects*
21	Social, Environmental and Economic Sustainability in the Context of Melanesian Mining Projects	Mining and Cultural Loss: Assessing and Mitigating Impacts in Papua New Guinea	Martha Macintyre marthaam@unimelb.edu.au Simon Foale simon.foale@jcu.edu.au
22	Support of Indigenous Knowledge for the Use and Conservation of Biological Diversity of Ethnic Minorities in Three Ecological Regions in Yunnan, Southwest China	Indigenous Knowledge, Biodiversity Conservation, and Poverty Alleviation Among Ethnic Minorities in Yunnan, China	Xu Jianchu J.C.Xu@cgiar.org
23	Tado Cultural Ecology Conservation Program	Countering the Loss of Knowledge, Practices and Species on Flores Island, Indonesia	Jeanine Pfeiffer jmpfeiffer@mindspring.com Elizabeth Gish, Tado and Waerebo Communities www.ecosea.org
24	Territorial Management in Brazil's Xingu Indigenous Park	A 'Life Plan' for the Park: Culturally Appropriate Management in Brazil's Xingu Indigenous Park	Darron Collins darron.collins@wwfus.org www.amazonteam.org
25	The Significance of Non-Timber Forest Products Utilization and Cultural Practices in Rural and Urban Households: Implications for Biocultural Diversity	Wild Resources and Cultural Values: Implications for Biocultural Diversity in South Africa	Michelle Cocks m.cocks@ru.ac.za
26	The use of Aboriginal Traditional Knowledge in Species Assessment: A Case Study of Northern Canada Wolverines	Aboriginal Traditional Knowledge and Assessment of Species at Risk: A Case Study from Northern Canada	Nathan Cardinal nathan.cardinal@pc.gc.ca
27	Traditional Ecological Knowledge Relating to Marine Environment and Fishing on Lihir	Countering Fish Stock Depletion through Traditional Knowledge, Tenure and Use of Marine Resources in Papua New Guinea	Martha Macintyre marthaam@unimelb.edu.au Simon Foale simon.foale@jcu.edu.au
28	Training Program of Indigenous Agro-Forestry Agents of Acre	Training Indigenous Agro-Forestry Agents in Acre, Brazil: Indigenous and Modern Technologies for Sustainability	Giulia Pedone giuliapedone@yahoo.it Renato Gavazzi renato@cpiacre.org.br
29	Whitefeather Forest Initiative	Combining Environmental Stewardship and Economic Renewal in Northern Canada: The Whitefeather Forest Initiative	Andrew Chapeskie Alex Peters Whitefeather Forest Initiative Whitefeather Forest Management Corporation Pikangikum First Nation Pikangikum, ON POV 2L0 www.whitefeatherforest.com
30	Worlds of Difference	Worlds of Difference: Local Culture in a Global Age	Jonathan Miller jon@homelands.org www.homelands.org/worlds

#	*Project names as given by contributors*	*Project descriptive titles as given in Chapter 4*	*Contact information for contributors and projects*
31	Wik, Wik-Way and Kugu Ethnobiology Project	Integrating Local and Scientific Knowledge: The Wik, Wik-Way and Kugu Ethnobiology Project in Queensland, Australia	Sarah Edwards sarah.edwards@pharmacy.ac.uk
32	Plant Resources: Traditional Knowledge of Irulars of Northern Tamil Nadu	Local Knowledge and Self-Determination for Conservation: The Case of the Irular of Tamil Nadu, India	C. Manjula manjula_c6@yahoo.co.in
33	Local Level Ecosystem Assessment in India	Recording Traditional Knowledge of Biodiversity for the People's Biodiversity Register of India	Yogesh Gokhale ssopan@yahoo.com
34	Medicinal Plants of Antiquity	Ancient Botanical Knowledge as Living Knowledge: Medicinal Plants of Antiquity Program	Alain Touwaide ewmedicinalplants@hotmail.com
35	Ethnocartography and Self-Demarcation of Indigenous Peoples' Lands in Venezuela as Tools for Biocultural Diversity	Tools for Biocultural Diversity Conservation: Community Mapping of Indigenous Peoples' Traditional Lands in Venezuela	Stanford Zent srzent@gmail.com
36	Weavers of Life (*Tejedores de Vida*)	Tejedores de Vida: Revitalizing Indigenous Identity and Nature-Based Knowledge in a Muisca Community, Colombia	Gabriel R. Nemogá grnemogas@gmail.com Carlos Mamanché (deceased)
37	Community-based Documentation of Indigenous Knowledge, Awareness and Conservation of Cultural and Genetic Diversity of Bottle Gourd (*Lagenaria siceraria*) in Kitui District in Kenya	Countering Local Knowledge Loss and Landrace Extinction in Kenya: The Case of the Bottle Gourd (*Lagenaria siceraria*)	Yasuyuki Morimoto y.morimoto@cgiar.org
38	Transforming the Cage	Recovering the Connection Between People and the Environment Through Ancestral Law in British Columbia, Canada	Patricia Vickers pjvickers@mac.com
39	*Ndee bini' bida'ilzaahi*: Pictures of Apache Land	Learning that Wisdom Sits in Places: Apache Students Reconnecting to Land and Identity in Arizona, US	Jonathan Long jwlong@fs.fed.us
40	Vanishing Voices of the Great Andamanese	Endangered Languages, Endangered Knowledge: Vanishing Voices of the Great Andamanese of India	Anvita Abbi anvitaabbi@hotmail.com www.andamanese.net/dictionary.htm
41	Knowledge and Language Revitalization in Hawaii	Teaching and Learning from an Indigenous Perspective: Knowledge and Language Revitalization in Hawaii	Chad Kālepa Baybayan kalepa_b@leoki.uhh.hawaii.edu

#	*Project names as given by contributors*	*Project descriptive titles as given in Chapter 4*	*Contact information for contributors and projects*
42	Mapping Aboriginal Cultural Values in the Wet Tropics World Heritage Area	Putting Australian Aboriginal Cultural Values on the Map: The Wet Tropics World Heritage Area as a Biocultural Landscape	Bruce White bruceanthro@yahoo.com
43	A Collaborative Social and Biological Study with Gamo Elders of the Importance for Biocultural Diversity of Living Indigenous Sacred Sites in the Gamo Gofa Montagnard Region of Southwest Ethiopia	Indigenous Sacred Sites and Biocultural Diversity: A Case Study from Southwestern Ethiopia	Desalegn Desissa desissa@yahoo.co.uk
44	Talking the Walk: Language as the Missing Ingredient of Biodiversity Conservation? An Investigation of Plant Knowledge in the West Usambara Mountains, Tanzania	Talking the Walk in Tanzania: Language as the Missing Ingredient of Biodiversity Conservation?	Samantha Ross S.Ross@uea.ac.uk
45	Ethnobotany of Indigenous People of the Southern Rift Valley and Southwestern Ethiopia	Biodiversity Conservation through Traditional Practices in Southwestern Ethiopia, a Hotspot of Biocultural Diversity	Zerihun Woldu zerihunw@bio.aau.edu.et

Appendix 4

Directory of Selected Resources on Biocultural Diversity

Compiled by Ellen Woodley

A directory of other selected resources on biocultural diversity, with a short description of each one, where possible, is presented here. The directory includes organizations, institutions and foundations that take a biocultural approach in their activities or that incorporate an attention to the links between biodiversity and cultural diversity in some of their programmes.

African Resource Centre for Indigenous Knowledge (ARCIK)
Professor Adedotun Phillips, Director (Correspondent)
Nigerian Institute of Social and Economic Research (NISER)
PMB 5 – UI Post Office
Ibadan, Nigeria
Tel: +234-22-400500
Fax: +234-02-8101194
Email: arcik@niser.org.ng

ARCIK is dedicated to multidisciplinary research and documentation activities in Africa's indigenous knowledge (IK) systems, which is localized knowledge unique to particular African societies and groups that has been institutionalized and passed through many generations up to the present. ARCIK concerns itself with the search, retrieval, storage and dissemination of information on IK systems in the social, economic, political, cultural and technological life of African societies. The centre provides bibliographic support to researchers in the area of IK. Additionally, it organizes and encourages IK research by staff and by other scholars in Nigeria and Africa. It also organizes conferences, seminars, workshops and symposia on various aspects of Nigerian and African IK systems.

Amazon Conservation Team (ACT)
www.amazonteam.org/

All of ACT's programmes have a profound interest in and commitment to the preservation of culture, biodiversity and health. Some of the ways that these goals are achieved are through the institution of Shamans and Apprentices programmes, support for comprehensive participatory mapping projects, and the establishment of traditional clinics.

Anthropology News (American Anthropological Association)
www.aaanet.org/

Assembly of First Nations (AFN)
www.afn.ca/

The Assembly of First Nations is the national organization representing First Nations citizens in Canada. The AFN represents all citizens regardless of age, gender or place of residence.

BirdLife International
www.birdlife.org

By focusing on birds, which are are excellent flagships and vital environmental indicators, and the sites and habitats on which they depend, the BirdLife Partnership is working to improve the quality of life for birds, for biodiversity and for people.

Cameroon Indigenous Knowledge Organisation (CIKO)
Professor C. N. Ngwasiri, Director (Correspondent)
PO Box 8437, Yaoundé, Cameroon
Tel: +237-322 181
Fax: +237-322 181 / 430 813
Email: ngwasiri@camnet.cm

CIKO is an independent, non-profit and non-partisan action-research and advocacy organization. Some of the objectives are: to promote the utilization of indigenous knowledge to enhance the development of Cameroon's agriculture, animal breeding, industry, commerce, education, health and culture; to organize seminars, workshops, conferences, training sessions, radio and television interviews for the propagation and sharing of indigenous knowledge.

Center for Biological Diversity
www.biologicaldiversity.org/swcbd/aboutus/index.html

The centre's ideology is based on the links between the health and vigour of human societies and the integrity and wildness of the natural environment. Beyond this extraordinary intrinsic value, animals and plants, in their distinctness and variety, offer irreplaceable emotional and physical benefits to our lives and play an integral part in culture. Their loss, which parallels the loss of diversity within and among human civilizations, causes impoverishment beyond repair.

Center for Indigenous Knowledge for Agriculture and Rural Development (CIKARD)
www.ciesin.org/IC/cikard/CIKARD.html

CIKARD's activities and current programmes are based on the following objectives: to act as a global clearinghouse for collecting, documenting and disseminating information on indigenous knowledge of agriculture, natural resource management, and rural development; to develop cost-effective and reliable methodologies for recording indigenous knowledge; to conduct training programmes and design materials on indigenous knowledge for extension and other development workers; to conduct interdisciplinary research on indigenous knowledge systems; to promote the establishment of regional and national indigenous knowledge resource centres; and to formulate agricultural and natural resource management policies and design technical assistance programmes based on indigenous knowledge. There are several country-based centres, such as: Bangladesh Resource Centre for Indigenous Knowledge (BARCIK); Burkina Faso Resource Centre for Indigenous Knowledge (BURCIK), Centre for Indigenous Knowledge Fourah Bay College (CIKFAB); Maasai Resource Centre for Indigenous Knowledge (MARECIK-tz); Sri Lanka Resource Centre for Indigenous Knowledge (SLARCIK); and Yoruba Resource Centre for Indigenous Knowledge (YORCIK).

Center for World Indigenous Studies (CWIS)
www.cwis.org

CWIS is an independent, non-profit research and education organization in the US dedicated to a wider understanding and appreciation of the ideas and knowledge of indigenous peoples and their social, economic and political realities. The centre fosters better understanding between peoples through the publication and distribution of literature written and voiced by leading contributors from Fourth World Nations. An important goal of CWIS is to establish cooperation between nations and to democratize international relations between nations and between nations and states. The World Council of Indigenous Peoples (WCIP) is a part of CWIS.

Centre for Cosmovisions and Indigenous Knowledge (CECIK)
Dr David Millar, Director (Correspondent)
cecik@africaonline.com.gh

CECIK's vision is for cosmovision-based endogenous development (development embedded in the indigenous knowledge, the spirituality and astrology of the people), to grow and become sustainable in Northern Ghana. CECIK supports development in which the communities themselves become the experts, who own and control the pace of development of sustainable livelihoods.

Centre for Indigenous Knowledge on Population Resource and Environmental Management (CIKPREM)
Professor D. S. Obikeze (Correspondent)
epseelon@aol.com

CIKPREM pursues the general objectives of promoting, retrieving, documenting, disseminating and integrating indigenous knowledge in its three special areas: population, resources and environment. It does this through research, conferences, publications and collaboration with people involved in the field.

Centre of Indian Knowledge Systems
www.ciks.org/

CIKS is developing a programme to strengthen and revitalize varied aspects of Indian knowledge systems. The CIKS approach is based on the premise that these strengths should become the basis on which today's needs and requirements can be met. The CIKS methodology involves taking an in-depth look at these ancient knowledge systems to gain a strong understanding of their workings and rationale, which is then used to develop solutions that are practical and feasible in today's context. CIKS strongly believes that the future lies in understanding and harnessing the potential of indigenous knowledge systems and integrating them into the mainstream of scientific, industrial and everyday thinking.

Centre for Tropical Alternative Agriculture and Sustainable Development (CATADI)
University of The Andes, Núcleo 'Rafael Range'
Apartado Postal # 22
Trujillo 3102, Estado Trujillo
Venezuela
Tel: +58-72-721672
Fax: +58-72-362177
Email: consuelo@cantv.net Dr Consuelo Quiroz, Coordinator/Correspondent

Centro di documentazione sui popoli minacciati
(Centre for Documentation on Threatened Peoples)
www.popoliminacciati@ines.org

The Christensen Fund
www.christensenfund.org/

The Christensen Fund (TCF) recognizes the interdependence of cultural and biological integrity and focuses its efforts on that component of diversity which has recently been coined as *biocultural* – namely the weave of humankind and nature, cultural pluralism and ecological integrity. The Fund aims to buttress the efforts of people and institutions who believe in a biodiverse world through place-based work in regions chosen for their potential to withstand and recover from the global erosion of diversity. TCF focuses on backing the efforts of locally recognized community custodians of this heritage, and their alliances with scholars, artists, advocates and others. As well, TCF fund international efforts to build global understanding of these issues.

Circle of Stories
www.pbs.org/circleofstories/storytellers/

Worldwide, the preservation of biological diversity is inextricably related to the preservation of cultural diversity. For native peoples, culture and environment are deeply interwoven. They are one and the same because everything comes from the Earth, and the land is often where ancestors reside. Certain plants, animals and land forms are religious symbols, sources of food and healing materials, and characters in myths and stories. When the land loses its nature and the plants and animals that enliven it, the stories and the songs live in shadow, and ways of life disappear. But in some locations, land and culture are being reclaimed and revitalized. All around the world, as species and cultures are driven to the edge of extinction, we are finally learning the relationship between cultural and biological diversity. If we protect places we must also protect the rich cultures and their knowledge. Native peoples lose their cultural foundation as the Earth suffers from contamination and exploitation, whether from coal, oil, gold and uranium mines, or nuclear waste dumps and weapons testing. At the same time, there are many exciting efforts to revitalize culture and reclaim ancestral lands. Master–apprentice language programmes match language bearers with youth eager to learn their language and ways; video and audio ethnography is helping to teach new generations traditional arts and sciences; land acquisition projects are establishing cultural and ecological preserves where the land is protected and tended by ceremony.

Compas Network
www.compasnet.org/

COMPAS (Comparing and supporting endogenous development) is an international network implementing field programmes to develop, test and improve endogenous development (ED) methodologies. Endogenous development is based on local peoples' own criteria of development, and takes into account the material, social and spiritual well-being of peoples. The COMPAS programme is coordinated by the ETC Foundation in The Netherlands. Compas produces a six-monthly magazine that attempts to stimulate development agencies and individuals to consider indigenous knowledge in the support of endogenous development. The magazine aims to be a forum of intercultural dialogue, promoting exchange on testing field methods, on-farm research and participatory approaches based on farmers' own concepts, institutions and cosmovision in the domains of agriculture, health and natural resources. By stimulating intercultural dialogue, indigenous institutions can be strengthened and enable communities to re-enforce their position locally, regionally and internationally. Community members can reverse the process of cultural erosion, aggravated by globalization, and actively experiment with combinations of ancient knowledge and new knowledge.

Conservation International
www.conservation.org

Degraded landscapes and dwindling species numbers spell tragic consequences because the loss of biodiversity reduces the quality of life for all. For indigenous peoples that depend on healthy and productive ecosystems to meet their daily needs, their very survival is at stake. We must protect the diversity of life, not only for its intrinsic value, but also because a vibrant, healthy society depends on our continued success in safeguarding our threatened natural assets.

Convention on Biological Diversity (CBD)
Article 8j of the CBD and Focal Area 5 of the CBD's 2010 Target explicitly acknowledge the important contribution that traditional knowledge makes to the conservation and sustainable use of biological diversity.

Cultural Conservancy
www.nativeland.org/who2.html

The Cultural Conservancy is a Native American non-profit organization dedicated to the preservation and revitalization of indigenous cultures and their ancestral lands. As a research, education, and advocacy organization, the conservancy provides mediation, legal information referral and audio recording services as well as educational programmes and materials and technical training on Native land conservation and land rights, cultural and ecological restoration, and traditional indigenous arts and spiritual values. The conservancy acknowledges the sacred relationship of Native peoples to the land and the essential role of Native peoples in preserving environmental integrity and biological diversity. It recognizes and supports the link between cultural and biological diversity and the principle of Native self-determination. The conservancy is committed to cross-cultural interaction for environmental problem solving, networking and peacemaking.

Cultural Survival
www.cs.org/

Cultural Survival partners with indigenous peoples to secure their rights to their lands, resources, languages, cultures and to promote their participation in the political life of the countries in which they live.

Dispatches from the Vanishing World
www.dispatchesfromthevanishingworld.com/

Dispatches from the Vanishing World is a forum for documenting and raising consciousness about the world's fast-disappearing biological and cultural diversity. It provides first-hand, in-depth reporting from the last relatively pristine places on Earth, identifies who and what is destroying them, and who is engaged in the heroic and often life-threatening struggle to save them. It provides foundations involved in environmental or cultural preservation with two services: (1) a full, independent assessment of their programme or cause, and (2) publicity by adapting the assessment for publication in one of the top American magazines or as a book.

EANTH-L (Ecological Anthropology Listserv)
www.eanth.org/index.php
eanth-l@listserv.uga.edu

This is a website and listserv for anthropologists interested in ecology, the environment and environmentalism. It is part of the American Anthropological Association, the professional society of American anthropologists.

Earth Island Institute
www.earthisland.org/

Life on Earth is imperilled by human degradation of the biosphere. The Earth Island Institute develops and supports projects that counteract threats to the biological and cultural diversity that sustain the environment. Through education and activism, these projects promote the conservation, preservation and restoration of the Earth.

European Centre for Nature Conservation (ECNC)
www.ecnc.nl/

The Nature and Society Programme seeks to understand and explore the interrelated processes of social and ecological structures. Society has an impact on the environment and therefore conservation policy aims to influence society's impact and the way that it interacts with nature and biodiversity. ECNC actively promotes, by bridging the gap between science and policy, the conservation of nature and especially of biodiversity in Europe, because of their intrinsic values and their relevance to the economy and European culture; thereby ECNC seeks the integration of nature conservation into other policies. ECNC's vision takes into account the interaction between ecosystems, the role of landscapes, the integration of nature considerations in economy, and the perception and appreciation of nature in the minds of the people.

Fauna and Flora International
www.fauna-flora.org/

Fauna and Flora International is working to address the threats facing the variety of life on Earth. Its vision is of a sustainable future for the planet, where biodiversity is effectively conserved by the people who live closest to it, supported by the global community.

Forests and Oceans for the Future
www.ecoknow.ca/

Forests and Oceans for the Future is a research group based at the University of British Columbia (BC) in Canada, that focuses on ecological knowledge research conducted in collaboration with north coast BC communities. The research is intended to help incorporate core community values and aboriginal and non-aboriginal knowledge in local sustainable forest and natural resource management. Research focuses on the following three key activities: (1) applied research into local ecological knowledge; (2) policy development and evaluation; and (3) educational materials designed to facilitate mutual respect, effective communication and knowledge-sharing between First Nations and other natural resource stakeholders.

Gaia Foundation
www.gaiafoundation.org

The Gaia Foundation is an international non-governmental organization (NGO) based in London. Its mission is the protection of cultural and biological diversity, and democracy. Gaia works in Amazonia, which is one of the greatest areas of cultural and biological diversity in the world. Gaia has formed an alliance with indigenous groups in Colombia through the COAMA (Consolidation of the Amazon Region) Program. The COAMA program won the Right Livelihood Award for 'Vision and work forming an essential contribution to making life more whole, healing our planet and uplifting humanity'. Gaia has also been working on a programme of activities to support African negotiators taking on the challenge of developing common positions at international forums such as the CBD and the World Trade Organization, to protect biological diversity and people's democratic control of their lives and resources.

The Global Diversity Foundation
www.globaldiversity.org.uk

The Global Diversity Foundation is concerned about the future of biodiversity, the languages people speak and the ways they interact with their cultural landscapes, in the belief that globalization can go hand in hand with diversity. But it requires education, research and sheer hard work in the form of long-term, community-based projects. The foundation works with local people to restore and conserve diverse traditions through research and education on biocultural diversity; and in the field it supports projects that improve the health, education and rights of communities under threat from the globalized economy.

Indigenous Environmental Network (IEN)
www.ienearth.org/

IEN is a network of indigenous peoples empowering indigenous nations and communities towards sustainable livelihoods, demanding environmental justice and maintaining the Sacred Fire of traditions.

Indigenous Knowledge and People Network
www.ikap-mmsea.org/

Indigenous and tribal communities and peoples should determine and participate fully in their country's development on the basis of their own indigenous knowledge (IK). Mutual support and cross-border relationships exist between indigenous and tribal peoples and communities throughout the region in order to promote indigenous knowledge for sustainable livelihoods to strengthen community organizations and networks for the transition of IK to younger generations;

to establish contacts, facilitate exchange visits and join efforts for sustainable development; to build a network on capacity building (CB) to support IK and peoples for biodiversity conservation and endogenous development; to implement and support CB activities for indigenous peoples and facilitators; to provide advice and training to development workers and researchers (NGO, academics, international and state); to promote ethnic people's own research on IK and culture; and to develop advocacy to promote indigenous knowledge in MMSEA (Mainland Montane South East Asia).

Indigenous Peoples of Africa Co-coordinating Committee (IPACC)
www.ipacc.org.za/eng/default.asp

IPACC is a network of 150 indigenous peoples' organizations in 20 African countries, whose purpose is to coordinate African indigenous peoples' advocacy strategy and activities. IPACC promotes recognition of and respect for indigenous peoples in Africa; promotes participation of indigenous African peoples in United Nations' events and other international forums and strengthens the leadership and organizational capacity of indigenous civil society in Africa. IPACC conducts particular pilot projects related to the intergenerational transmission of traditional knowledge of biodiversity; the assessment and certification of traditional knowledge, competencies and skills; and innovative approaches to fighting poverty by using sustainable indigenous approaches to natural resources management. IPACC works in partnership with the Technical Centre for Agricultural Cooperation with Rural Areas (CTA EU-ACP); Cybertracker Foundation; African Biodiversity Network; Indigenous Information Network and UNESCO's working group on Education for Sustainable Development.

Indonesian Resource Center for Indigenous Knowledge (INRIK)
www.melsa.net.id/~inrik

INRIK was established in Padjadjaran University in 1992 to conduct activities that promote indigenous knowledge. The major aim of the centre is to obtain a clear understanding of indigenous knowledge and natural resource management as a basis for developing appropriate models for rural development strategies in order to improve existing practices or, at the very least, to prevent the further degradation of resource management systems in Indonesia. INRIK has links with global networks of professionals and institutions engaged in indigenous issues.

Interinstitutional Consortium for Indigenous Knowledge (ICIK)
http://icik-psu.com/index.php?p=1_18_Africa

ICIK, located in the College of Education at Pennsylvania State University, is part of a global network comprised of more than 20 indigenous knowledge resource centres in North and South America, Europe, Asia, Africa and Oceania. ICIK is the only currently active indigenous knowledge resource centre located in the US. ICIK is a network that promotes communication among community residents, students, university faculty and staff from across the Commonwealth of Pennsylvania who share an interest in diverse local knowledge systems and would like to engage with communities that generate locally useful knowledge to enable their survival in a rapidly globalizing society. ICIK maintains a listserv, a website and a resource library; sponsors seminars, conferences and workshops; and produces an indigenous knowledge book series. ICIK also encourages collaborative research that addresses issues of community scholarship and transformation of the academy to embrace two-way communication with local communities.

The International BioPark Foundation
www.biopark.org/

The International BioPark Foundation is dedicated to re-establishing the natural balance that best supports the life of our planet by honouring the interdependent nature of the relationships of all life forms and re-educating humanity to the unique responsibility that we share in this endeavour.

International Centre for Integrated Mountain Development
www.icimod.org

ICIMOD, together with its partners and regional member countries, is committed to a shared vision of prosperous and secure mountain communities committed to peace, equity and environmental sustainability. ICIMOD's mission is to develop and provide integrated and innovative solutions, in cooperation with national, regional and international partners, which foster action and change for overcoming mountain people's economic, social and physical vulnerability.

International Indian Treaty Council (IITC)
www.treatycouncil.org

The IITC is an organization of indigenous peoples from North, Central and South America and the Pacific working for the sovereignty and self-determination of indigenous peoples and the recognition and protection of indigenous rights, traditional cultures and sacred lands. IITC's objectives are to seek, promote and build the official participation of indigenous peoples in the United Nations and its specialized agencies, as well as other international forums; to seek international recognition for treaties and agreements between indigenous peoples and nation-states; to support the human rights, self-determination and sovereignty of indigenous peoples; to oppose colonialism in all its forms, and its effects upon indigenous peoples; to build solidarity and relationships of mutual support among indigenous peoples of the world; to disseminate information about indigenous peoples' human rights issues, struggles, concerns and perspectives; and to establish and maintain one or more organizational offices to carry out IITC's information dissemination, networking and human rights programmes.

International Indigenous Forum on Biodiversity (IIFB)
www.iifb.net

The IIFB was formed during the Third Conference of the Parties to the CBD (COP 3) in Buenos Aires, Argentina, in November 1996. The IIFB is a collection of representatives from indigenous governments, indigenous non-governmental organizations and indigenous scholars and activists that organize around the CBD and other important international environmental meetings to help coordinate indigenous strategies at these meetings, provide advice to the government parties, and influence the interpretations of government obligations to recognize and respect indigenous rights to the knowledge and resources.

International Network on Ethnoforestry (INEF)
http://tech.groups.yahoo.com/group/inef/

INEF is a peer group of concerned foresters, scientists, international agencies and NGOs working for the documentation, dissemination and integration of indigenous knowledge on forest management with formal forestry, in various cultures and with indigenous peoples across the globe. Context-specific knowledge helps INEF to address various questions on forest management and can help society to overcome the crisis of habitat destruction and over-exploitation. The Indian Institute of Forest Management, Bhopal, India is the centre for INEF.

International Society of Ethnobiology (ISE)
http://ise.arts.ubc.ca/
For two decades, ISE has actively promoted and supported the inextricable linkages between biological and cultural diversity and the vital role of indigenous and local peoples in stewardship of biological diversity and cultural heritage, which includes recognition of land and resource rights, as well as rights and responsibilities over tangible and intangible cultural and intellectual properties. The ISE is committed to understanding the complex relationships that exist between human societies and their environments. A core value of the ISE is the recognition of indigenous peoples as critical players in the conservation of biological, cultural and linguistic diversity. The vision of the ISE is reflected in its Code of Ethics, to which all Members are bound.

International Union of Anthropological and Ethnological Sciences (IUAES)
www.wcaanet.org/member/iuaes
The aim of IUAES is to enhance exchange and communication among scholars of all regions of the world, in a collective effort to expand human knowledge. In this way it hopes to contribute to a better understanding of human society, and to a sustainable future based on harmony between nature and culture.

The International Union for the Conservation of Nature (IUCN)
www.iucn.org

The policy of IUCN's Commission on Environmental, Economic and Social Policy (CEESP) (www.iucn.org/about/union/commissions/ceesp) is to provide insights and expertise on ways to harmonize biodiversity conservation with the crucial socioeconomic and cultural concerns of human communities, such as livelihoods, poverty eradication, development, equity, human rights, cultural identity, security and the fair and effective governance of natural resources. CEESP's Theme on Culture and Conservation focuses on the importance of incorporating culture and cultural diversity into IUCN's policy and programme. Together with IUCN's World Commission on Protected Areas (WCPA), CEESP set up the Theme on Indigenous and Local Communities, Equity, and Protected Areas (TILCEPA) and the Theme on Governance, Equity and Rights (TGER). TGER has a Task Force on the Cultural and Spiritual Values of Protected Areas. The WCPA has also played an important role in bringing together and disseminating methodologies for the identification and quantification of the economic values of protected areas. This work is complemented by that of the Task Force, which seeks to identify, define and provide guidelines for managing the cultural and spiritual dimensions of protected areas. CEESP has recently taken on the Indigenous and Community Conserved Areas (ICCA) initiative (www.iucn.org/about/union/commissions/ceesp/topics/governance/icca/ceesp_icca_database/). These are sites, resources or species that are voluntarily conserved through community knowledge, values, practices, rules and institutions.

Inuit Tapiriit Kanatami (ITK)
www.itk.ca/

The Environment Department at ITK is dedicated to protecting and advancing the place of Canada's Inuit in the use and management of the Arctic environment and its resources. Despite the considerable changes that have occurred in our society over the past 50 years, the relationship between Inuit and the land continues to be a fundamental element of Inuit culture and identity. ITK is dedicated to ensuring that the Arctic environment and its resources are protected and managed properly. ITK communicates regularly with the appropriate departments of the regional Inuit organizations to keep them informed of national and international initiatives while seeking direction from them when ITK is developing a plan of action.

Language and Ecology
www.ecoling.net/

Language and Ecology is an online journal focusing on critical analysis of discourses implicated in environmental destruction, and exploration of alternative discourses and their potential to contribute to ecological sustainability. The journal also publishes articles that explore the application of ecolinguistics to education for sustainable development.

Mountain Forum
www.mtnforum.org/resources/library/diva97a.htm

Mountain Forum is a global network of individuals and organizations concerned with the well-being of mountain people, their environments and cultures. Mountain Forum seeks to bring lessons and experiences of mountain people into policy discussions at national and international levels with the aim of improving their livelihoods and promoting the conservation of mountain environments and cultures.

Nature Conservation Research Centre (NCRC)
www.ncrc-ghana.org/

The Nature Conservation Research Centre (NCRC) is a Ghanaian non-profit, private voluntary organization implementing conservation initiatives to promote a greater awareness and protection of the natural, historic and cultural diversity of Ghana and ultimately the West African sub-region. NCRC endorses a core philosophy that conservation in Ghana must be pursued in settings where there are cultural and economic incentives for its implementation. Conservation in Ghana should emerge from local cultural belief systems, and must have tangible economic returns for humans living in the area. Without culture and economics as core elements, we believe conservation efforts will not succeed in this country. In line with this core philosophy, NCRC seeks to use positive cultural practices and income-generation potential to advance its goals and objectives.

Ogiek
www.ogiek.org

The Ogiek are indigenous peoples living mainly in Kenya's Mau and Mt Elgon Forests, who are fighting to remain in their ancestral homeland. The former government tried to force them out of the forests, allegedly to protect the environment. The Ogiek pose no environmental threat, but instead are actually the guardians of these forests since time immemorial.

OneWorld International Foundation
www.oneworld.net/

The OneWorld network and portal brings together the latest news, action, campaigns and organizations in human rights and global issues across five continents and in 11 different languages, published across its international site, regional editions, and thematic channels. Many of these are produced from the South to widen the participation of the world's poorest and marginalized peoples in the global debate.

Open Forum on Participatory Geographic Information Systems and Technologies (PPGIS)
http://ppgis.iapad.org/

PPGIS is an electronic forum on the participatory use of geo-spatial information systems and technologies. Three distinct discussion lists serve as global avenues for discussing issues, sharing experiences and good practices related to community mapping, Participatory GIS (PGIS), Public

Participation GIS (PPGIS) and other geo-spatial information technologies. These technologies are used in participatory settings to support integrated conservation and development, sustainable natural resource management and customary property rights in developing countries and First Nations. Participatory GIS developed out of participatory approaches to planning and spatial information and communication management and is the result of a spontaneous merger of Participatory Learning and Action. A good PGIS practice is embedded into long-lasting spatial decision-making processes, is flexible, adapts to different socio-cultural and biophysical environments, depends on multidisciplinary facilitation and skills, and builds essentially on visual language. The practice integrates several tools and methods while often relying on the combination of 'expert' skills with socially differentiated local knowledge. It promotes interactive participation of stakeholders in generating and managing spatial information and it uses information about specific landscapes to facilitate broad-based decision-making processes that support effective communication and community advocacy. By placing control of access and use of culturally sensitive spatial information in the hands of those who generated them, PGIS practice could protect traditional knowledge and wisdom from external exploitation.

A Plan to Protect Bio-diversity and Indigenous Culture in Sri Lanka
http://vedda.org/bio-diversity_plan.htm

Sri Lanka's indigenous 'first people', the Veddas or Wanniyalaeto ('forest-dwellers') as they call themselves, have inhabited Sri Lanka's semi-evergreen monsoon dry forest, the Wanni, for at least 16,000 years. To this day, their detailed knowledge of their habitat, including its fauna and flora, remains unsurpassed. Development activities in the 20th century, however, have drastically reduced both the Wanniyalaeto people and their traditional forest habitat to the extent that unless measures are taken soon, not only many species of fauna and flora but also the indigenous human culture that successfully managed the forest environment for millennia face almost certain extinction. More recently, however, the Sri Lanka government's policy towards its indigenous citizens and their role in the development process has undergone changes reflecting a more sympathetic perception of indigenous aspirations. In particular, with the growing recognition of a precipitous drop in the island's forest cover and related adverse effects upon wildlife and general fertility due to reduced rainfall, the government is now more inclined to avail itself of Wanniyalaeto expertise in protecting the remaining forest cover and wildlife. A final window of opportunity to preserve biodiversity and indigenous culture simultaneously now presents itself, but for a short time only before it is too late. A plan is now being formulated by the NGO Cultural Survival of Sri Lanka in consultation with the Wanniyalaeto that will eventually return the day-to-day management of the Maduru Oya National Park back to the Wanniyalaeto with the active cooperation and participation of government ministries and international development aid agencies.

Rural Research Centre Iran (RRC)
Contact: Dr Mohammed H. Emadi, Deputy Head
Email: rrciri@neda.net
www.wiserearth.org/organization/view/2da58cac43d77da2de81e4ad9a944c45

The Ministry of Agriculture and Rural Development in Iran is involved with systems agriculture and rural development, systems thinking and its application in agriculture, indigenous knowledge and rural studies with the focus in Iran and Asia. In addition, agricultural extension and education, participatory resource management and development planning, and participatory methodologies are used.

Russian Association of Indigenous Peoples of the North, Siberia and the Far East (RAIPON)
www.raipon.net/

RAIPON was created in 1990 at the First Congress of Indigenous Peoples of the North. The Association now represents 41 indigenous groups whose total population is around 250,000 people. RAIPON is a public organization that has as its goal the protection of human rights and the defence of the legal interests of indigenous peoples of the North, Siberia and the Far East. It assists in finding solutions to environmental, social and economic problems, and the problems of cultural development and education. RAIPON works to guarantee the rights to the protection of native homelands and traditional ways of life as well as the right to self-governance according to the national and international legal standards.

Saami Council
www.saamicouncil.net/

The Saami Council, founded in 1956, is a voluntary Saami non-governmental organization, with Saami member organizations in Finland, Russia, Norway and Sweden. The primary aim of the Saami Council is the promotion of Saami rights and interests in these four countries where the Saami live, to ensure affinity among the Saami people, to attain recognition for the Saami as a nation and to maintain the economic, social and cultural rights of the Saami in the legislation of the four states.

Society for Applied Anthropology
www.sfaa.net/

The Society has for its object the promotion of interdisciplinary scientific investigation of the principles controlling the relations of human beings to one another, and the encouragement of the wide application of these principles to practical problems.

Society for the Study of Indigenous Languages of the Americas (SSILA)
www.ssila.org/

SSILA was founded in December 1981 as the international scholarly organization representing American Indian linguistics, and was incorporated in 1997. Membership in SSILA is open to all those who are interested in the scientific study of the languages of the native peoples of North, Central and South America. The Society has approximately 900 members, more than a third of them residing outside the US.

South African San Institute (SASI)
www.sasi.org.za

SASI is an independent, non-governmental organization that mobilizes resources for the benefit of the San peoples of southern Africa as mandated by the Working Group of Indigenous Minorities in Southern Africa (WIMSA) and other San organizations. This is done through activities such as community mobilization, fund raising, lobbying, networking, training, building strategic alliances and capacity building on issues related to culture, language, income generation, health and social environment, and land rights. The goal is for the San peoples of southern Africa to achieve permanent control over their lives, resources and destiny.

Spirit of the Sage Council
www.sagecouncil.com/

The council is an all volunteer grassroots non-profit project and coalition of American Indians, environmental organizations, citizens' action groups, scientists, legal experts and wildlife advocates, dedicated to the protection and conservation of America's natural and cultural resources. The number of members and coalition support groups is, on average, 1000 individual members and 30 groups throughout the US, Mexico and Canada. The council addresses both the biological and cultural significance of conserving the habitats of rare, threatened and endangered species. An ecosystem is considered complete when the indigenous peoples of the land are present and actively involved in natural resource management. The biocentric philosophy enables the council to be able to reach consensus quickly by taking a position, and actions, that benefit the Earth and flora and fauna first.

Survival International
www.survival-international.org/

Survival International is the only international organization supporting tribal peoples worldwide. It was founded in 1969 after an article by Norman Lewis in the UK's *Sunday Times* highlighted the massacres, land thefts and genocide taking place in Brazilian Amazonia. Today, Survival has supporters in 82 countries. It works for tribal peoples' rights in three complementary ways: education, advocacy and campaigns. Tribal people themselves are offered a platform to address the world. Survival works closely with local indigenous organizations, and focuses on tribal peoples who have the most to lose, usually those most recently in contact with the outside world. Educational programmes are aimed at people in the 'West' or 'North' in order to destroy the myth that tribal peoples are relics, destined to perish through 'progress'. Survival promotes respect for their cultures and the contemporary relevance of their way of life. Survival also plays a major role in ensuring that humanitarian, self-help, educational and medical projects with tribal peoples receive proper funding. A good example is the Yanomami medical fund, which succeeded in virtually eliminating malaria in some Indian areas.

Tebtebba Foundation
www.tebtebba.org

Tebtebba (Indigenous Peoples' International Centre for Policy Research and Education) is an indigenous peoples' organization born out of the need for heightened advocacy to have the rights of indigenous peoples recognized, respected and protected worldwide. Established in 1996, and based in the Philippines, Tebtebba seeks to promote a better understanding of the world's indigenous peoples, their worldviews, their issues and concerns. In this effort, it strives to bring indigenous peoples together to take the lead in policy advocacy and campaigns on all issues affecting them. The organization's vision is to have a world where indigenous knowledge and indigenous peoples' rights are respected and protected by all nations and societies; where there are unified yet diverse and vibrant indigenous peoples' movements at the local and global levels which enhance the self-determination and sustainable development of indigenous peoples and their territories. The mission is to be an indigenous peoples' organization and a research, education, policy advocacy and resource centre working with indigenous peoples at all levels and arenas. Tebtebba seeks the recognition, promotion and protection of indigenous peoples' rights and aspirations while building unities to uphold social and environmental justice and sustainability. This is to be achieved by reinforcing the capacities of indigenous peoples for advocacy, campaigns and networking; research, education, training and institutional development; and by actively articulating and projecting indigenous peoples' views and perspectives.

United Nations (UN) Declaration on the Rights of Indigenous Peoples
www.un.org/esa/socdev/unpfii/en/declaration.html

The UN Declaration on the Rights of Indigenous Peoples, approved in 2007, states that 'control by indigenous peoples over developments affecting them and their lands, territories and resources will enable them to maintain and strengthen their institutions, cultures and traditions, and to promote their development in accordance with their aspirations and needs'. It also recognizes that 'respect for indigenous knowledge, cultures and traditional practices contributes to sustainable and equitable development and proper management of the environment'.

United Nations Millennium Declaration
www.un.org/millennium/declaration/ares552e.htm
The declaration recognizes the importance of the diversity of belief, culture and language, and affirms that societal differences should be cherished as precious assets of humanity.

United Nations Development Programme (UNDP)
www.undp.org/

UNDP's perspective is that the cultural dimension of human life is playing an increasing role in the definition of human development and human well-being.

United Nations Educational, Scientific and Cultural Organisation (UNESCO)
http://portal.unesco.org/en/ev.php-URL_ID=29008&URL_DO=DO_TOPIC&
URL_SECTION=201.html

UNESCO adopted the Universal Declaration on Cultural Diversity in 2001 and the Convention on the Protection and Promotion of the Diversity of Cultural Expressions in 2005. UNESCO's Endangered Languages Programme focuses on safeguarding the world's linguistic heritage, while its LINKS (Local and Indigenous Knowledge Systems in a Global Society) programme focuses on the strengthening and revitalization of traditional knowledge. An initiative on science and traditional knowledge was carried out by the International Council for Science (ICSU, 2002), following up on some of the outcomes of the UNESCO World Conference on Science (UNESCO, 2000). UNESCO also has a Main Line of Action on Biodiversity and Cultural Diversity, and the Programme on Man and the Biosphere (MAB) recognizes that traditional forms of land use often conserve ancient breeds of livestock and crop landraces.

United Nations Environment Programme (UNEP)
www.unep.org/

UNEP complemented its Global Biodiversity Assessment (Heywood, 1995) with an extensive review of the cultural and spiritual values of biodiversity (Posey, 1999).

Uruguayan Resource Centre for Indigenous Knowledge (URURCIK)
Pedro de Hegedüs, Coordinator (Correspondent)
CEDESUR
PO Box 20.201
Sayago, Montevideo 12.900
Uruguay
Tel/fax: +5-982-308 16 03
Email: phegedus@adinet.com.uy

Venezuelan Resource Secretariat for Indigenous Knowledge (VERSIK)
Dr Consuelo Quiroz, National Coordinator (Correspondent)
Centre for Tropical Alternative Agriculture and Sustainable Development (CATADI)
University of The Andes, Núcleo 'Rafael Range'
Apartado Postal # 22
Trujillo 3102, Estado Trujillo
Venezuela
Tel: +58-72-721672
Fax: +58-72-362177
Email: consuelo@cantv.net Dr Consuelo Quiroz, Coordinator/Correspondent

WiserEarth
www.wiserearth.org

WiserEarth is an online network that helps the global movement of people and organizations working toward social justice, indigenous rights, and environmental stewardship connect, collaborate, share knowledge, and build alliances. Among its working groups, WiserEarth has established a Working Group on Biocultural Diversity, found at www.wiserearth.org/group/biocultural_diversity, where professionals and community members involved in biocultural diversity work share experiences, text and multimedia resources, as well as thoughts and dialog through online forums.

Working Group for Indigenous Minorities in Southern Africa (WIMSA)
www.san.org.za/wimsa/home.htm

WIMSA was established in 1996 at the request of the San in South Africa, Botswana, Namibia, Zambia and Zimbabwe, to provide a platform for their communities to express their problems, needs and concerns. WIMSA is required to advocate and lobby for San rights, to establish a network for information exchange among San communities and other concerned parties, and to provide training and advice to San communities on tourism, integrated development projects and land tenure. One of the objectives is to support the San in regaining their identity and pride in their cultures, thereby improving their self-esteem.

World Resources Institute (WRI): Conserving Cultural Diversity
www.wri.org/publication/content/8215

WRI's publication *Keeping Options Alive: The Scientific Basis for the Conservation of Biodiversity* contains a chapter about conserving cultural diversity, which focuses on the threats to indigenous peoples, their knowledge of biodiversity, and their territorial rights.

World Social Forum (WSF)
www.forumsocialmundial.org.br/home.asp

The World Social Forum is an open meeting place where social movements, networks, NGOs and other civil society organizations opposed to neoliberalism and a world dominated by capital or by any form of imperialism come together to share ideas, to debate ideas and to network for effective action. The World Social Forum is also characterized by plurality and diversity.

Worlds of Difference
http://homelands.org/worlds/

Worlds of Difference uses radio documentaries to explore the impact of global change on traditional societies worldwide. The goal of the project is to stimulate public discussion on questions of diversity, tradition, identity and change. Most **Worlds of Difference** stories are intimate, sound-rich documentary features that bring listeners into the homes and communities of people facing critical decisions about their changing ways of life. The project includes 40 stories from 27 different countries and addresses a theme: the relationship between culture and the market; culture and language; culture and place; culture and religion; culture and the past; and culture and the path to the future. This project grows out of an urgent concern for the rights and welfare of cultural groups whose worlds are changing as a result of forces beyond their control. The project recognizes that these processes are complex and the choice of stories, as well as the commitment to the highest journalistic standards, reflect this.

World Wide Fund for Nature (WWF)
www.wwf.org

WWF has taken a cultural approach to biodiversity conservation in such work as 'Indigenous and Traditional Peoples of the World and Ecoregion Conservation: An Integrated Approach to Conserving the World's Biological and Cultural Diversity', WWF-International and Terralingua, Gland, Switzerland (available at http://assets.panda.org/downloads/EGinG200rep.pdf). More recently, WWF, Equilibrium and the University of Birmingham, UK, published a paper 'Food Stores: Using Protected Areas to Secure Crop Genetic Diversity' (available at http://assets.panda.org/downloads/food_stores.pdf).

Appendix 5
About Terralingua

Terralingua (www.terralingua.org) is an international non-governmental organization (NGO) whose mission is to support the integrated protection, maintenance and restoration of the biocultural diversity of life – the world's invaluable heritage of biological, cultural and linguistic diversity – through an innovative programme of research, education, policy-relevant work and on-the-ground action. Terralingua pioneered the field of biocultural diversity, which was launched by the conference 'Endangered Languages, Endangered Knowledge, Endangered Environments' organized by Terralingua in Berkeley, California in 1996 (see L. Maffi (ed.), *On Biocultural Diversity*, Smithsonian Institution Press, 2001).

Since its founding in 1996, Terralingua has developed a comprehensive programme of work, the Global Biocultural Diversity Assessment (GBCDA), which has received support from the Ford Foundation, The Christensen Fund and the International Development Research Centre (Canada) among others. The GBCDA has focused on:

- promoting understanding of biocultural diversity through research and education;
- mapping and analysing the global and regional distributions of biocultural diversity;
- developing integrated indicators to measure and monitor the global and sub-global state and trends of biocultural diversity;
- supporting the maintenance and restoration of biocultural diversity through field projects with local communities (currently in the Sierra Tarahumara of northern Mexico);
- fostering a community of practice in biocultural diversity conservation, through the development of this sourcebook and of a network of biocultural diversity conservation practitioners.

Terralingua has collaborated extensively with other international organizations as well as academic institutions, including the American Museum of Natural History, the CBD, Conservation International, the Field Museum of Natural History, IUCN, the Millennium Ecosystem Assessment, the Smithsonian Institution, the University of Florida, UNEP, UNESCO, WWF and many others. Through these partnerships, as well as by reaching a broad audience of researchers, practitioners, grassroots organizations and the general public through publications and educational activities, Terralingua has been instrumental in making biocultural diversity an object of academic enquiry and placing biocultural diversity on the international policy agenda. In April 2008, in partnership with IUCN and the American Museum of Natural History, Terralingua co-organized the symposium 'Sustaining Cultural and Biological Diversity in a Rapidly Changing World: Lessons for Global Policy', held in New York, which yielded policy-relevant inputs on biocultural diversity in preparation for IUCN's Fourth World Conservation Congress (Barcelona, October 2008). The Congress hosted a week-long Biocultural Diversity and Indigenous Peoples Journey, to which Terralingua contributed several events. A resolution on 'Integrating Culture and Cultural Diversity into IUCN's Policy and Programme', co-sponsored by Terralingua along with Center for Biodiversity and Conservation, American Museum of Natural History and Macquarie University Centre for Environmental Law, was approved by the IUCN Members Assembly at the Congress. The interrelation of cultural diversity with biodiversity is now recognized in IUCN's 2009–2012 Programme of Work, as well as in recent UNEP and UNESCO documents.

Appendix 6
About the Authors

Luisa Maffi, PhD, is co-founder and Director of Terralingua. She is one of the originators of the concept of biocultural diversity. Her edited volume *On Biocultural Diversity: Linking Language, Knowledge, and the Environment*, published by Smithsonian Institution Press (2001), is widely regarded as a foundational work in this field. Other key publications include: the co-authored chapter 'Linguistic Diversity' in the volume *Cultural and Spiritual Values of Biodiversity* edited by Darrell Posey (ITP/UNEP, 1999), the co-edited book *Ethnobotany and Conservation of Biocultural Diversity* (New York Botanical Garden Press, 2004), the co-edited special issue of IUCN's *Policy Matters* 'History, Culture, and Conservation' (vol 13, 2004), the review essay: 'Linguistic, cultural, and biological diversity' (*Annual Review of Anthropology*, vol 34, 2005), and the co-authored section on 'Culture' in the Biodiversity chapter of UNEP's 4th Global Environment Outlook Report (GEO-4, UNEP, 2007). Dr Maffi has a background in linguistics (BA, University of Rome) and anthropology (PhD, University of California at Berkeley), with research experience in Africa (Somalia) and Mesoamerica (Mexico), which was supported by the US National Science Foundation and US National Institutes for Health. She has acted as consultant and collaborator with several international organizations (WWF, UNEP, UNESCO, IUCN, the World Bank) and has contributed to international processes such as the World Summit on Sustainable Development, the Millennium Ecosystem Assessment, the Biodiversity Indicators Partnership, the Convention on Biological Diversity, UNEP's GEO-4 report and IUCN's Theme on Culture and Conservation.

Ellen Woodley, PhD, has been involved with environment and development issues for 20 years. Her interdisciplinary PhD in Rural Studies integrated the environmental and social sciences to effectively understand and use local ecological knowledge for ecosystem management and conservation in the Solomon Islands. This followed an MSc that involved research on agrobiodiversity in Sulawesi, Indonesia. Her international research includes direct involvement with small communities on issues relating to biodiversity conservation, cultural diversity, smallholder agriculture and food security in the South Pacific (Papua New Guinea and Solomon Islands), West Africa (Sierra Leone and Mali), Indonesia (Sulawesi) and with First Nations in Canada. Her international research also includes work at the science–policy interface, as a coordinating lead author for the 2005 Millennium Ecosystem Assessment and the 2007 UNEP GEO-4. As a consultant to the UN FAO and the IITC, Woodley conducted research on the development of cultural indicators of indigenous peoples' food and agroecological systems, which support and develop capacity for a rights-based approach to the revitalization of traditional practices that conserve biodiversity and provide for the sustainable use of biodiversity. She was also a research intern with the International Development and Research Center (IDRC) in the Sustainable Use of Biodiversity Program. Ellen has conducted baseline vegetation studies and ecosystem classification in Canada and Papua New Guinea. Woodley's long-standing collaboration with Terralingua has been as coordinator of the Global Sourcebook on Biocultural Diversity. She is a member of IUCN's CEESP and the Commission on Ecosystem Management (CEM), as well as of the Canadian Sustainability Indicators Network (CSIN), the social science working group (SSWG) of the Society for Conservation Biology (SCB) and the Indigenous Cooperative on the Environment (ICE).

Index

Page numbers in bold refer to illustrations.

Abbi, Anvita 51–53
Aboriginal Rainforest Council (ARC) 66
Adams, W.M. 194, 195
adat ceremonial language (Indonesia) 56
Addis Ababa University 32, 146
Africa
  biocultural diversity of Central Africa 6
African projects 28–42
African violets 36
Agta people (Philippines) 60, 135
alcoholism 53
alien species 132
Amazon Basin, biocultural diversity 6
Amazon Conservation Team (ACT) 100, 118–119, 146
American Museum of Natural History (AMNH) 19–20
Andamanese tribes (India) 51–53, 142, 152, 162, 168
Angkol, Agustinus **55**
anthopology 3, 9, 15–16
Anthropocene 15
antiquity 43–44, 143
Apache language 91, 162
Apache people (US) 90–92, 138, 148, 149
apartheid 41
Arctic projects 78–82
Asian projects 45–60
Association of Kom Forest Management Institutions (Cameroon) 29
Association of Oku Forest Management Institutions (Cameroon) 29
Aswani, Shankar 71–73
ATIX (Brazil) 118
Aurukun ethnobiology (Australia) 64–65, 144, 148–149, 164
Australia
  Aurukun ethnobiology 64–65, 144, 148–149, 164
  community participation 144
  digital gap 63
  edible gum *(mardiwa)* **150**
  languages 61–62, 64–65, 67, 142, 144, 149, 162
  native land rights 67
  Northern Australian Aborigines 63, 66–67
  recognition of indigenous knowledge 136
  transmission methods 148–149
  Western Australian Aboriginal environmental knowledge transmission 61–62, 149
  Wet Tropics World Heritage Area (Queensland) 66–67, 138–139, 144–145, 151, 157
  Wik, Wik-Way and Kigu Aborigines (Queensland) 64–5, 144, 148–149, 164
Australian National University 70
Australian Research Council (ARC) 68
Avicenna 43
Awa people (Ecuador) 111
Ayurveda traditional medicine (India) 49

Balée, William 120
Bambuti people (Uganda) **124**
Bamenda Highlands forest project (Cameroon) 28–29, 136, 140, 149, 151, 157, 168
Bannerman's Turaco 29
Barnard, Jonathan 28–29
Basso, K. 90, 91, 92
Baybayan, Chad Kālepa 77

beliefs *See* customary beliefs
Bikini Atoll (Marshall Islands) 76
biocultural diversity
  biodiversity and 4
  causes of loss 130–135
    economic development 131
    environmental degradation 131
    industrial agriculture 134
    insecurity of tenure 131, 139
    socio-economic factors 8, 131, 132–133
    sources of pressure on ecosystems 132
  collaboration and conservation 144–147
  conditions for success 158–168
  cultural affirmation 136–138
  defining 4–6
  entry points 129, 135–136
  global and regional correlations 6–8
  indicators 26, 176
  international policies 179, 180–181
  linguistic diversity and 6–7
  linkages 135–144
  local level 9–10, 169
  multi-disciplinary origins 3
  organic concept 3
  policy advances and gaps 179–184
  policy implementation 151–153
  population density and 8
  project policies 25
  projects *See* projects
  recommendations *See* recommendations
  relevance 13–20
  research gaps 175–178
  reviving cultural knowledge 138–140
  reviving languages 141–144
  sustainability and 18–20, 170
  transmission methods 147–151
  trends 10–11
biodiversity
  biocultural diversity and 4
  causes of loss 17–18
  comparative thesaurus 123, 136, 142–143, 151
  Convention on Biological Diversity *See* Biodiversity Convention
  cultural affirmation and 136–138
Biodiversity International 34
biology 13
Birdlife International 29
Boruca people (Costa Rica) 99
bottle gourds *(kitete)* 33–35, 140, 141, 142, 148, 161, 163, 164, 166, 168
Brazil
  Amazon policy 121–122
  ATIX 118
  capacity building 167
  community participation 146, 165
  Ka'apor reserve, protection 120
  Kayapó people **14**
  land rights 118–119, 139, 160–161
  language transmission 149
  logging 120
  mapping indigenous lands 118–119, 139
  National Indian Foundation (FUNAI) 117, 118, 120
  public policy 168
  Terra Indigena Alto Tunaçu 120
  training indigenous agro-forestry agents in Acre 121–122, 139, 153, 160, 168
  Xingu Indigenous Park 117–119, 137, 139, 144, 146, 149, 159, 160–161, 165
Bribri people (Costa Rica) 99
Brunca people (Costa Rica) 99
Brundtland Report 182, 191
busy lizzies 36

Cabécar people (Costa Rica) 99
caci (whip dance) 56
calabashes 33–34
Cameroon
  Bamenda Highlands forest project 28–29, 136, 140, 146, 149, 151, 157, 168
  CBD and 29, 157
  community participation 29, 146
  ecotourism 167
  forest management 159, 164
  government support 168
  Kilum-Ijim forest 28–29
  traditional medicine 140
Canada
  ancestral law in British Columbia 87, 140
  capacity building 167
  community participation 84, 145, 146, 165
  COSEWIC 81
  endangered species and traditional knowledge 81–82
  First Nations, well-being and culture 9–10
  Gitxaała land use 85–86, 139, 149, 158

Gwich'in place names 78–80, 130, 138, 152, 158, 162, 165, 167
indigenous languages 142
International Development Research Centre 94
knowledge transmission 150, 162
National Historic Sites 80
Northern Boreal Initiative 84, 152
Pikangikum First Nation 83–84, 138, 142, 146, 151, 163
policy development 152–153
protected areas 84
Species At Risk Act 81, 152, 164
sustainable land use 78–80, 163
traditional health practices 88–89, 159
Whitefeather Forest Initiative 83–84, 138, 142, 145, 146, 150, 151, 152–153, 158, 162, 163, **164**
wolverines 81–82, 137, 152, 157
Canadian Institute of Health Research (CIHR) 88, 89
capacity building 166–167
Cardinal, Nathan 81–82
Caribbean 27
Carrera, Rigoberto 99–100
Catholicism 56, 140
CBD *See* Convention on Biological Diversity
Chachi people (Ecuador) 111
Chapeskie, Andrew 83–84
Charles Darwin University (Australia) 63
Chemonics International 111
China
deforestation 45–46
economic reforms 45–46
indigenous knowledge revival 46
knowledge transmission failures 159
poverty alleviation 45, 46
Yunnan ethnic minorities 45–47, 151, 159
Chorotega people (Costa Rica) 99
Christensen Fund, The 32, 94
Cibecue Community School (US) 90–92, 162
climate change 15, 17, 74, 75, 131, 181
Cocks, Michelle 41–42
Cofan people (Ecuador) 111
collaboration *See* participatory approach
Collins, Darron 117–119
Colombia
cultural identity 161
Los Hijos del Maíz (The Chidren of Corn) 103–104
lack of government support 168
land rights 160, 161, 179
Muisca community 103–106, 160, 161, 165, 168
Tejedores de Vida (Weavers of Life) 103–106
colonialism 39, 54, 103, 133, 162
commercial agriculture 113, 134
communitarianism 113
community of practice 184–186
community participation *See* participatory approach
Commonwealth Scientific and Industrial Research Organization 70
comparative approaches 176–177
conservation biology 3
Conservation Foundation 54
Convention on Biological Diversity
biocultural diversity and 179, 180
Cameroon and 29, 157
CUBIC 2000 46
IPRs and traditional knowledge 49, 153
Programme of Work on Protected Areas 72, 187
project guidelines 151
COPROALDE 101–102
COSEWIC 81
Costa Rica
capacity building 99–100, 167
community participation 147
indigenous groups 99–100
industrial agriculture 134
land rights 100, 139
national policy changes 152
Ngäbere language 99–100, 142
oral history 100, 142, 150
seed exchange 101–102, 137, 149
Couples for Christ (CFC) 73
crocodiles 59–60, 135, 137, 149, 163, 167–168, 169
culture
See also biocultural diversity
cultural affirmation 136–138
cultural blind spots 19
cultural diversity, meaning 4
sustainability and 182
Cultures and Biodiversity Congress (CUBIC 2000) 46
customary beliefs
See also sacred sites

Andaman Islands (India) 52
anthropological perspective 16
Australian Aborigines 64
building on 156
Colombia 106
conservation and 18, 25
Ethiopia 30–31, 149
Irular people (India) 54
Kenya 141
Marshall Islands 74–75
Nepalese food security and 48
Peru 114–115, 149
Philippines 59, 135, 137
reviving 138–140
taboos *See* taboos

'Dance for the Earth and for her Peoples' 124–125, 145–146, 151
Dar es Salaam University 37
Dassanetch peoples 32
DeMarco, John 28
Desissa, Desalegn 30–31
Devanagari script 53
development
See also sustainability
definition for indigenous people 183
economic growth 192
loss of biocultural diversity and 131
digital technology 63
Dizi peoples 32
Ducula oceanica (Micronesian Pigeon) 76

EARTh 123, 136, 142–143, 151
Easter Island 17
ecological anthropology 9
ecology 3
economic growth 192
ecopolitics 18
ECOSEA 56–58, 144
ecotourism 37, 57–58, 65, 105, 110, 163, 167
Ecuador
CAIMAN project 111, 163
capacity building 110–111, 167
community participation **110,** 145
indigenous land rights 111, 137, 139, 160, 179
indigenous peoples 110–111, 127, **178**
policy guidelines 151
edible gum *(mardiwa)* **150**

education needs 187–189
Edwards, Sara 64–65
El Dorado 104
emigration *See* migration
endangered languages *See* languages
endangered species *See* crocodiles; wolverines
Eñepa people (Venezuela) 107–109, 140, 148, 167
English language, Tanzania 38, 162
environmental degradation, loss of biocultural diversity and 131
essentialism 18
Ethiopia
community participation 32, 145, 146
Gamo peoples 30–1, 140, 144, 145, 149, 158, 161–162
lack of government support 168
sacred sites 30–1, 32, 130, 138, 140, 161–162
traditional practices and conservation 32, 138
Ethiopian Wildlife and Natural History Society 31
ethnobiology 3, 9, 10
Ethnobotanical Conservation Organization for Southeast Asia (ECOSEA) 56–58, 144
ethnoecology 3
ethnosphere 4, 193
European projects 43–44

Farmers' Union of Catalonia 104
fishing
Canadian FIrst Nations **83,** 84, **85**
Papua New Guinea 70, 137, 153
Philippines 59
Solomon Islands 72
Flores Island (Indonesia) 55–58
Foale, Simon 68–69, 70
folk songs *See* songs
Food and Agriculture Organization (FAO) 181
food security 48, 73, 89, 102, 105, 130, 132, 182
forest management
See also logging
Amazon 119
Brazil 122, 152
Cameroon 28–29, 136, 146, 149, 159, 164

Chinese deforestation 45–46
Gitxaała people (Canada) 86
India 54
Pikangikum people (Canada) 84
sacred forests of Ethiopia 31, 32
Friends of Gamo Gofa Sacred Sites Association 31
Friends of Usambara 37
Fundación CRUSA 100
Fundación TUVA 99
Fundecooperación 100

Galen 43
Gamo peoples (Ethiopia) 30–1, 140, 144, 145, 149, 158, 161–162
Gavazzi, Renato 121–122
gender *See* women
Gesellschaft für Technische Zusammenarbeit (GTZ) 46
Gish, Elizabeth 55–58
Gitxaała people (British Columbia) 85–86, 139, 149, 158
Global Consultation on the Right to Food and Food Security for Indigenous Peoples 182
globalization 4, 37, 64, 134, 157, 182
Gokhale, Yogesh 49–50
Govan, Hugh 99–100
government support 167–168
green revolution 59
Greenheck, Felipe Montoya 101–102
GTZ (Gesellschaft für Technische Zusammenarbeit) 46
Guaja people (Brazil) 120
Guatuso people (Costa Rica) 99
Gwich'in peoples (Canada) 78–80, 130, 138, 152, 162, 165, 167
Gwich'in Social and Cultural Institute (GSCI) 78, 79, 158, 165

Halcyion cloris (collared kingfisher) 58
Hamar people (Ethiopia) 32, 138
Hans Rausing Endangered Language Fund 52
Hardison, Preston 18
Hawaii
indigenous knowledge and language 77, 142
languages 77, 142, 148
transmission methods 148
herbal medicine *See* traditional medicine
Los Hijos del Maíz (The Chidren of Corn) 103–104
Hill, Sekagya Yahaya 39–40
Hippocrates 43
Homeland Productions (US) 126–127
Hotï people (Venezuela) 107–109, 140, 148, 167
Huaorani people (Ecuador) 110–111
Huetar people (Costa Rica) 99
human rights 111, 181, 183, 185
humans
humans-environment interactions 15–18
philosophy of separation from nature 13–15
hunting 17, 52, 54, 59, **78,** 111, 118, 131, 135, 139, 142

Ikpeng people (Brazil) 119
Ilocano language (Philippines) 60
immigration *See* migration
Impatiens 36
Index of Biocultural Divesity 6
Index of Linguistic Diversity 11
India
Adivasis 54
forest management 54
Great Andamanese 51–53, 142, 152, 162, 168
indigenous languages 51–53, 142, 147, 162
Irular of Tamil Nadu 54, 147, 165
knowledge transmission 147
lack of government support 168
National Biological Diversity Act (2002) 49, 153, 179
People's Biodiversity Register 49–50, 137–138, 153, 157, 164
Indigenous and Community Areas (ICCAs) 125, 187
indigenous knowledge
See also specific countries
approaches to 18–19
biocultural diversity and 3
biodiversity conservation and 32
encouraging 156
focus on 25
TEK *See* traditional environmental knowledge
indigenous languages *See* languages
indigenous people
See also specific countries; specific groups

approaches to 19
conservationism debate 17–18, 19
definition of development 183
ecopolitics 18
food security 182
romanticizing 18
sacred sites *See* sacred sites
UN Declaration on the Rights of Indigenous People 103, 181
Indonesia
community participation 144, 147, 165
ecotourism 57, 163
Flores Island 55–58
languages 56, 58
project principles 157
Tado Community 55, 138, 144, 147, 163, 165
Waerebo Community 55, 140, 144, 147, 165
industrial agriculture 113, 134
institutions, strength of local 158–160
intellectual property rights 49, 99, 153
intergenerational transmission 24, 162, 183–184
International Development Research Centre 94
International Indian Treaty Council 181
International Society of Ethnobiology, Code of Ethics 46, 57, 65, 157
International Union for the Conservation of Nature (IUCN) 19–20, 100, 124, 151, 179, 181, 193–194, 195
internet 127, 186
Irular people (India) 54, 147, 165
Irular Tribal Women's Welfare Society 54
Isabela State University (Philippines) 60
Ishizawa, Jorge 112–116
Italy
EARTh project 123
Institute for Atmospheric Pollution 123
medicinal plants of antiquity 43–44
IUCN 19–20, 100, 124, 151, 179, 181, 193–194, 195

Jaru language (Western Australia) 62, 142, 144, 149, 162
Jeanrenaud, S. J. 194, 195
Jero language 52–53 (india)
Johannesburg Declaration on Sustainable Development (2002) 179

Ka'apor people (Brazil) 120
Kalinga people (Philippines) 59–60, 135
Kamba people (Kenya) 33–35, 141
Kaxinawá people (Brazil) **121,** 121–122
Kayapó people (Brazil) **14**
Kempo Manggarai language (Indonesia) 56
Kenya
bottle gourds 33–35, 140, 141, 142, 161, 163, 164, 166, 168
community participation 166
government support 168
indigenous languages 142
National Museums of Kenya 34
keystone societies 15
Kichua people (Ecuador) 111
Kimberley Language Resource Centre Aboriginal Corporation 61–62, 144
Kimbugu language (Tanzania) 37, 38, 142
Kimoni, Jemima 35
kingfishers 58
Kisambaa language (Tanzania) 37, 38, 142
Kiswahili (Tanzania) 36, 37, 38, 162
kitete (bottle gourds) 33–35, 140, 141, 142, 148, 161, 163, 164, 166, 168
Konso people (Ethiopia) 32, 138
kraals 42
Kritsch, Ingrid 78–80
Kugu Aborigines (Queensland) 64–65, 144
Kuku Nyungkul Aborigines (Queensland) 67
Kyanika Adult Women's Group (KAWG) 34–35, 141

Ladino language 127
Lagenaria siceraria 33–35
land rights
Australia 67
Brazil 118–119, 139, 160–161
Colombia 160, 161, 179
Costa Rica 100, 139
Ecuador 111, 137, 139, 160, 179
loss of biocultural diversity and 131
Mexico 161
security of tenure 131, 139–40, 160–161, 183
Solomon Islands 71–73
supporting 156
Venezuela 107–109, 140, 160, 179
languages
Australia 61–62, 64–65, 67, 142, 144, 149, 162

biocultural diversity and 6, **7**
Brazil 149
Canada 142
causes of loss 133
Colombia 161
Costa Rica 99–100, 142
databases 63, 148, 149
effect of loss 19
endemic languages and higher vertebrates 6, **7**
environmental thesaurus 123, 136, 142–143, 151
global threats 11
Hawaii 77, 142, 148
importance of retention 9, 25
India 51–53, 142, 147
Indonesia 56, 58
Kenya 142
logosphere 4, 193
Marshall Islands 76, 148
Mexico 97, 142, 148
mother tongues as educational language 38
Nepal 48, 143–144
resources for nature 19
revitalization 127, 141–144, 156
Tanzania 36–38, 142, 148, 162
transmission methods 26, 147–151, 184
Ugandan traditional medicine and 39
UNESCO policies 180
Venezuela 142
vitality factors 10
Latin American projects 93–122
Leiden University 60
Lihir Gold Limited (LGL) 68–69
Lihir Island (Papua New Guinea) 68–69, 70, 137, 153
lingko 55
localism 9–10, 169
logging 28, 59, 87, 94, 95, 110, 120, **133,** 161
logosphere 4, 193
Long, Jonathan 90–92
López-Zent, Eglée 15–16
Luzon Island (Philippines) 59–60, 135, 167–168

MABUWAYA Foundation 60
Macintyre, Martha 68–69, 70
Madras University 54
Majuro Atoll (Marshall Islands) 76
Maleku people (Costa Rica) 99
Mamanché, Carlos 103–106
Mamanché, Rosa de 106
Manggarai peoples (Indonesia) 55–58, 140
Manjula, C. 54
Maori language 127
mardiwa (edible gum) **150**
Marsden, Dawn 88–89, 155–156
Marshall Islands
knowledge transmission 148
loss of languages 76, 148
marine protected areas 74
Pandanus tectorius 76
reconnecting with biocultural heritage 76
taboos and conservation 74–75
Marshall Islands Conservation Society 76
Marton-Lefèvre, Julia 193
Mazzocchi, Fulvio 123
medicine *See* traditional medicine
Me'en peoples 32
megafauna 17
Melbourne University 68, 70
Menzies, Charles 85–86
Mexico
capacity building 167
community participation 165
drinking water 95
eco-cultural health in Sierra Tarahumara 93–98, **133,** 138, 142, 144, 148, 149–50, 159, 165, 167
knowledge transmission 149–151
logging 94, 95, **133,** 161
migration 94
Rarámuri language 97, 142, 148
Rarámuri people 93–98, 144, **145, 160,** 161, 165
transmission methods 149–50, 159
Meya, Wilhelm 10
Micronesian Pigeon *(Ducula oceanica)* 76
migration
biocultural diversity and 9, 17, 131
Brazil 120
Mexico 94
original migration from Africa 51–52
Peru 115
Philippines 59, 135
Millennium Development Goals 177, 179
Millennium Ecosystem Assessment 195
Miller, Jonathan 126–127
MILPA Inc. 101

mining 60, 64, 68–69, 70, 85, 94, 110, 111, 131, 148
Mittermeier, Cristina 14
monocropping 101, 131, 134
monotheism 31, 132
Moose, Whitehead 83–84
Morimoto, Yasuyuki 33–35, 166
Mühlhaüsler, Peter 19
Muisca people (Colombia) 103–106, 160, 161, 165, 168
Mursi people (Ethiopia) 32
Museo do Indio (Brazil) 120
museums 34, 120, 148

National Museum of the American Indian (US) 120
National Museums of Kenya 34
natural disasters 132
nature
  humans as separate from 13–15
  humans-environment interactions 15–18
Nemogá, Gabriel 103–106
Nepal
  food security and traditional beliefs 48, 130, 140, 149
  languages 48, 143–144
  rice 48, 143–144, 158, 163
  selroti 48
Netherlands, Costa Rica and 100
networks 185–186, 187
Ngäbe-Buglé people (Costa Rica) 99–100, 139, 147, 152, 167
Nisga'a people (British Columbia) 87, 140
noosphere 4
North American projects 83–92
Norwegian University of Life Sciences 48
Nyerere, Julius 36

Occitan 127
oral history, Costa Rica 100, 142, 150
organic farming 101, 134
over-harvesting 70, 132
Ozies, Lydia **150**

Pacific
  megafauna 17
  projects 61–77
Palma, Luis and Tomás **93**
Panama, indigenous peoples' rights 179
Pandanus tectorius 76
Pant, Laxmi 48
Papua New Guinea
  fish stock depletion 70, 137, 153
  Kinami Mountain **68**
  mining and cultural loss 68–69, 148
participatory approach
  Biodiversity Convention 46
  Cameroon 29, 146
  Canada 84, 145, 146, 165
  Ecuador **110,** 145
  establishing partnerships 165–166
  Ethiopia 32, 145, 146
  Indonesia 144, 147, 165
  Kenya 166
  Mexico 165
  Nepalese project 48
  Peru 113–114
  project analysis 144–147
  selection criterion 24
  Solomon Islands **71**
  sustainability and 195
  Yunnan Initiative 47
Payesos Foundation 105
Paz, Octavio 127
Pedone, Giulia 121–122
Peru
  agrarian reform 113, 134
  biodiversity **16,** 112
  chacras 112, 114, 115
  community participation 113–114
  conservation incentives 163
  forms of knowledge **188**
  NACAS 113–115
  national policies 152
  potato farming **126,** 127, 134
  PRATEC 113–116, 134, 152
  rural education 112–116, 159–160
  traditional beliefs 114–115, 149
  transmission failures 159
Peters, Alex 83–84
Pfeiffer, Jeanine 55–58
Philippines
  acculturation processes 135
  crocodiles 59–60, 135, 137, 149, 163, 167–168, 169
  customary beliefs 59, 135, 137
  government support 167–168
  Indigenous People's Rights Act (1997) 179
  insurgencies 60

sustainability 163
transmission failures 159
philosophy, humans separate from nature 13–15
Pikangikum peoples (Canada) 83–84, 138, 142, 146, 151, 163
Ploeg, Jan van der 59–60, 169
policy implementation 151–153
population density 8
potato farming **126,** 127, 134
precautionary principle 178
project analysis
biodiversity through cultural affirmation 136–138
causes of biocultural loss 130–135
collaboration and conservation 144–147, 165–166
commonalities and differences 155–157
conditions for success 158–168
capacity building 166–167
collaborative partnerships 165–166
conclusions 169–171
fostering sustainability 163
government support 167–168
intergenerational transmission 162
planning with TEK 163–165
securing land tenure 160–161
strength of local institutions 158–160
strengthening cultural identity 161–162
entry points 129, 135–136
linkages 135–144
methodology 26
policy implementation 151–153
revival and traditional approaches 129–30
reviving cultural knowledge 138–140
reviving languages 141–144
transmission methods 147–151
projects
See also specific countries
Africa 28–42
analysis *See* project analysis
Arctic 78–82
Asia 45–60
biocultural diversity policies 25
choice of focus 25
conditions for success 158–168
customary beliefs and conservation 25
endogenous approach 24
Europe 43–44
global projects 123–127
guiding ideas 3
indigenous knowledge 25
integrative approaches 23–24
intergenerational transmission 24
language maintenance 25
Latin America 93–122
methodology 25–26
North America 83–92
overview 27–127
Pacific 61–77
participation *See* participatory approach
selection criteria 23–25
synergies 23–24
PROMETRA Uganda 39
protected areas
Canada 84
CBD and 72, 187
Marshall Islands 74
Solomon Islands 71–73, 137, 149, 152, 157, 163

Queensland Natural Resource Management 66
Queiroz, João de 110–111

radio 126–127
Ramos Rosa, Cristóbal 114–115
Rapport, David J. 93–98
Rarámuri language (Mexico) 97, 142, 148
Rarámuri people (Mexico) 93–98, 144, **145, 160,** 165
recommendations
community of practice 184–186
comparative approaches 176–177
education 187–189
networks 185–186, 187
policy advances and gaps 179–184
research gaps 175–178
synergies 186–187
value shift 189
religion
See also customary beliefs
Catholicism 56, 140
local languages and 37
monotheism 31, 132
Rengifo, Grimaldo 112–116
research gaps 175–178
resource scarcity 8
restructuring ecosystems 132
revival, traditional approaches and 129–30

ritual festivals 31, **33,** 48, 56, 115, 130, 140, 149, 158
Rodeo-Chediski wildfire 91
Ross, Samantha 36–38
Roviana Conservation Foundation 72
Roviana language (Solomon Islands) 73

sacred sites
  Apache people 90–92, 138
  Ethiopia 30–1, 32, 130, 138, 140, 161–162
  India 49
  Indonesia 56
  Muisca people (Colombia) 103
  taboos 140
  Xingu peoples (Brazil) 118
Saintpaulia 36
Sampi, Bonnie Maresha **61**
School of Oriental and African Studies (London) 52
Scurrah, Maria **126**
Secoya people (Ecuador) 111
security of tenure *See* land rights
seed exchange 101–102, 137, 149
self-determination 54, 139, 169, 179, 183
selroti 48
Smithonian Institution 44
societal implosion 17
Soliman the Magnificent **43**
Solomon Islands
  customary tenure systems 71–73
  marine protected areas 71–73, 137, 149, 152, 157, 163
songs 34, 52, 53, 56, 100, 140, 142, 144, 148
South Africa
  homelands 41
  kraals 42
  sustainability 163
  wild resources and cultural values 41–42, 138
  Xhosa women **41**
Spain, colonization of Latin America 103
Spanish Farmers in Solidarity 104
Strait Island 52–53
strategic essentialism 18
Stuart, Barbara **61**
Suggashie, George 84
sustainability
  biocultural approach 18–20
  Canadian land use 78–80, 85–86
  components 191
  fostering 163
  future 191–196
  humans-environment interactions 17–18
  lessons 170
  mantra 192
  recognition of culture for 182
  survey methodology 26
  Xingu Indigenous Park (Brazil) 119
Sustainable Livelihood Framework (UK) 182
synergies 23–24, 186–187

taboos 17, 30, 32, 59, 74–75, 135, 137, 140
Tado Community (Indonesia) 55, 138, 140, 144, 147, 163, 165
Tado Community Training and Research Centre 56
Tado Cultural Ecology Conservation Program 56
Tado Upland Rice Conservation Project 56
Tagalog language (Philippines) 60
TAMI 63
Tanzania
  Chake Chake village **196**
  herbal knowledge **143**
  languages 36–38, 142, 148, 152, 162
Tanzanian Forest Research Institute 37
technological fix 192
Teilhard de Chardin, Pierre 4
Tejedores de Vida (Weavers of Life) 103–106
Tembé people (Brazil) 120
Térraba people (Costa Rica) 99
Terrain Cultural Resource Management 67
Terralingua 94, 95, 96, 186
thesauri 123, 136, 142–143, 151
Thompson, Nick 92
Tibon, Jorelik 74–75, 76
Timbira people (Brazil) 120
Touwaide, Alain 43–44
traditional beliefs *See* customary beliefs
traditional environmental knowledge (TEK)
  approaches to 18–19
  erosion 10
  planning conservation with 163–165
  value 9–10
  VITEK 11
traditional medicine
  antiquity 43–44, 143
  Ayurveda system 49

  Cameroon 140
  Colombia 104, 105, 106
  Costa Rica 99–100
  Indonesia 56
  Irular people 54
  reviving knowledge 140
  Uganda 39–40, 130, 140, 159
tropical forests 8, 15
Turkey **43**
Turnbull, David 63

Uganda
  African spirituality 39–40, 140
  Bambuti community **124**
  languages 39
  traditional medicine 39–40, 130, 140, 159
UNEP 179, 180
UNESCO 11, 179, 180, 194
UNESCO World Heritage Convention 67, 151
United Kingdom, Sustainable Livelihood Framework 182
United Nations
  Declaration on the Rights of Indigenous Peoples 103, 168, 181
  Millennium Development Goals 177, 179
  Permanent Forum on Indigenous Issues and the Indigenous Global Caucus 27
  traditional knowledge and 180, 181
  Working Group on Indigenous Populations (UN-WGIP) 57
United States
  Apache language 91, 162
  Apache sites 90–92, 138, 148, 149
  Department of Agriculture 76
  National Museum of the American Indian 120
  National Park Service 76
  National Public Radio 127
  nuclear tests in Bikini Atoll 76
  USAID 111, 145
University of British Columbia 86, 88
University of Hawaii 77

Valera, Humberto 114
value shift 189
Velasco Alvarado, General Juan 113
Velde, Nancy Vander 74–75, 76
Venezuela
  capacity building 167
  indigenous language 142
  knowledge transmission 148
  lack of government support 168
  land rights 107–109, 140, 160, 179
  mapping traditional lands 107–109, 140
  policy guidelines 151
Verran, Helen 63
Vickers, Patricia 87
Vitality Index of Traditional Environmental Knowledge (VITEK) 11

Waerebo Community (Indonesia) 55, 140, 144, 147, 165
Waorani people (Ecuador) **178**
waste 132
Waurá people (Brazil) 119
Welsh langage 127
West Manggarai Swiss Australia Tourism Association (WiSATA) 57
Wet Tropics Management Authority 66
Wet Tropics World Heritage Area (Queensland) 66–67, 138–139, 144–145, 151, 157
whip dance *(caci)* 56
White, Bruce 66–67
Whitefeather Forest Initiative (Canada) 83–84, 138, 142, 145, 146, 150, 151, 152–153, 158, 162, 163, **164**
Wik Aborigines (Queensland) 64–65, 144
Wik-Way Aborigines (Queenland) 64–65, 144
Wild, Robert 124–125
Woja Conservation Area (Marshall Islands) 74
Woldu, Zerihun 32
World Summit on Sustainable Development (2002) 179
wolverines 81–82, 137, 152, 157
women
  Ethiopian biodiversity and 32
  Irular tribe 54
  Kyanika Adult Women's Group (Kenya) 34–35, 141
  literacy, Mexico 95, 97
  Manggarai women (Indonesia) 58
  research on role of gender 176
  Tanzanian focus groups **36**
  Xhosa women **41**
World Conservation Congress (2008) 181, 193
World Heritage Convention 67, 151
World Parks Congress (2003) 124, 187

'Worlds of Difference' radio series 126–127, 134

Xingu Indigenous Park (Brazil) 117–119, 137, 139, 144, 146, 149, 159, 160–161, 165
Xu Jianchu 45–47

Yasuni National Park (Ecuador) 110–111
Yunnan Initative (China) 45–47, 151, 159

Zápara people (Ecuador) 127
Zapotec language 127
Zent, Stanford 11, 15–16, 107–109